A SINGLE SWIM

In Collaboration with Swim Above Water
and the Courtney Nash Amoeba Awareness
Foundation Inc.

by Kristina Circelli

Editing by Juli's Elite Editing
Formatting by Donna Dull
Cover Art by Sarah Hansen/Okay Creations

ISBN-10:0976372878
ISBN-13:978-0-9763728-7-5

This book is a combination of facts about Courtney Nash's life and other stories as told by family members, as well as scientific input by medical specialists. Some names have been changed or shortened by request, or to protect the individual, as indicated by a * symbol.

First Edition
1 2 3 4 5 6 7 8 9 10

For Courtney
And for Those Who Lost Their Lives, and the Families That Survive Them

"When I count my blessings, I count my horse twice."
- Courtney Nash

From a Mother's Memory
A Poem by P.J. Nash-Ryder

"As of right now my life is in God's hands." -
Courtney Nash

I'd like to tell you a little story about this sixteen-year-old girl,
Who has changed the world.
Her name was Courtney Jean.
She was sweet; she was caring, always lending a hand.
She loved her brother, her horses; she was a big Taylor Swift fan.
She loved to fish and play in the mud.
She had a big heart and gave her love.
She thought of others so they could live.
Her organs she donated - what more can you give.
She gave of herself so others could live.
I now sit here at the river, out at Hatbill Park,
Knowing that you are still here,
Way way up high.
There you are, the angel in the sky.
Still watching us as we grow,
To make sure that we all know,
God has taken you home to a safe and happy place.
No worries, no fear,
For the Lord says he is near.
Be happy for me.
I am finally free.
We will be together again one day,
And you can show us the way.
This story will continue as we keep in our hearts,
The example of one life.
One sixteen-year-old girl with one big heart.

Table of Contents

Foreword

Two years ago, I received a phone call that a cousin of mine was ill. Though we weren't close, hers was a name I'd heard often, spoken through the family grapevine as words often pass. I knew her family, knew how tight knit they were, closely resembling my own immediate family.

That phone call spoke as much of my cousin as it did her family. My mother told me the story, as much as anyone knew at that time — Courtney was in the hospital with an illness no one could identify, and it wasn't looking good. "I just wanted to let you know," I remember my mom telling me, "so you could keep Courtney and P.J. [her mom] in your thoughts."

Only a couple days later Courtney passed away, and it was then that we learned about the deadly amoeba that had taken her life — an amoeba none of us had ever heard about, despite growing up and living in an area that allows for a perfect environmental host. In the days and months that followed Courtney's death, we learned more about a silent killer that lurked around us our entire lives, never having known about the brain-eating parasite that could have taken any one of our lives on any given day.

Later, P.J., Courtney's mother, would start a foundation dedicated to her daughter and to raising awareness to a single-cell killer that has been taking lives since the 1960s — perhaps even earlier, though these are the earliest on medical record. Personally, I kept up with P.J.'s foundation and many benefits via Facebook and word of mouth, admiring and appreciating the way she kept her daughter's memory alive in such a positive way.

When the topic of writing a book about Courtney was first brought up (by someone not involved with the foundation), I shied away from the idea, worried I would not be able to do this beautiful teenager justice. A fiction writer through and through, I feared that attempting a true story rooted in such scientific fact would be beyond my capabilities. But, the more I watched P.J. honor her daughter, the more I wanted to honor P.J., and everything she found the strength to do in the wake of Courtney's passing.

A Single Swim began as a tribute to Courtney, to tell her story and highlight P.J.'s awareness organization. Throughout the course of researching the amoeba and writing the book, it became much more. In addition to working with the Courtney Nash Amoeba Awareness Foundation in Florida, I also met up with the members of Minnesota's Swim Above Water, all of whom shared their stories with me and gave me greater insight into the devastating effects the amoeba not only has on its victims, but on the families left behind.

One by one I heard the stories from parents who lost children, wives who lost husbands, families who lost a brother or son, and one by one I realized that A Single Swim was no longer about one person or one death. Instead, it became a book about anyone, everyone, who has ever swam in a lake, water skied in the river, slid down a ride at a local water park. It no longer mattered that I wasn't a scientist or even a nonfiction author. All that mattered was that I told their stories, that I told them well, and that I helped these people raise awareness.

A Single Swim is a testament to all of the victims of *Naegleria fowleri*, to all of the families left behind, to everyone who has dedicated their lives to raising awareness, and to the medical and scientific professionals who took the time to share their knowledge and experiences.

To this end, I want to thank everyone who has been a part of this journey.

First and foremost, I want to thank P.J. Nash-Ryder for allowing me the opportunity to write A Single Swim. I brought the idea to her before putting the first word to paper, requesting permission to tell her daughter's story. Her enthusiasm for the novel and the time spent sending me information, mementos of Courtney, and simple words of encouragement and gratitude helped transform this work from a mere book into a celebration of one girl's life.

Wanting the book to be as best it could be, P.J. also put me in contact with Swim Above Water, a northern-US organization that has played a pivotal role in the creation of A Single Swim. I want to extend my thanks to Heidi LaMeyer and Jim Ariola. Heidi has been a huge support for the book, and is one of the most enthusiastic and pleasant people I've ever had the pleasure of meeting. With her help, A Single Swim became a collection of heartfelt stories set within the context of scientific awareness. Heidi has been an integral part of A Single Swim and I look forward to working with her, as well as those involved with Swim Above Water, in the future.

I also want to thank everyone who took the time to speak with me, or gave permission to use their loved one's story. I am not a parent, and so I cannot imagine the heartache these gracious and courageous mothers and fathers have gone through. But, as each one spoke with me about their child, I heard not only the grief in losing their son or daughter, but also the love that was still so very much at the forefront of their hearts. I thank each and every parent who stands up to speak about their loved one, keeping their memories alive, being so willing to speak about those traumatic days in order to bring awareness to the world.

I want to offer a word of thanks to the scientists and medical specialists who offered their knowledge on this deadly amoeba. First, to Dr. Sandra Gompf, MD and Infections Disease Sub-Specialist. Dr. Gompf provided valuable information on the amoeba as well as shared the story of her son, who passed away of this deadly amoeba. Dr. Mobeen Rathore, MD, CPE, FAAP, FIDSA, FACPE, provided a unique viewpoint on the amoeba and environmental factors, utilizing his expertise in the area. And finally, Lisa*, infectious disease physician, and Bob Walton, biological and environmental sciences specialist, provided additional comments and insights for the manuscript, bringing both a medical and biological approach to the book.

Lastly, I want to say thank you to those who cannot hear my words, but whom I hope will hear me in spirit. It has been an honor getting to know those who lost their lives to this deadly amoeba, learning about these unique and unforgettable individuals, and getting to tell their stories. I wish that fate had taken a different direction for each and every one of you, allowing you to live the life so eagerly awaited. I hope that A Single Swim does its part to honor you as much as my life has been touched by your stories.

Chapter 1

An Introduction

On August 13th, 2011, surrounded by loved ones, Courtney Jean Nash passed away after a sudden and fatal attack.

It was a mother's worst nightmare and the most difficult decision to make — to take her sixteen-year-old daughter off life support, allowing her spirit to pass on. The decision followed several days of worrying and praying for a medical miracle, and finally forced to accept a death that was not to be avenged. For Courtney's passing was not a common case or an easily treated illness; her death was the cause of a killer seen only by the eyes of a microscope.

Naegleria fowleri.

For years this silent killer, an amoeba that invades the brain and causes a fatal infection to nearly everyone who comes in contact with it, passed by relatively unknown, residing in the warm waters of stagnant rivers and lakes throughout the United States, as well as other parts of the world where temperatures often run higher. For Titusville, Florida teen Courtney Nash and her family, it was a killer that didn't have a name until that fateful sunny day in Southern Florida.

In the two years that followed her death, her family would work tirelessly to help ensure no other mothers, fathers, sisters, brothers, wives, husbands, are never forced to make that same decision. A Single Swim follows their efforts, and shares the stories of others who have lost loved ones to this deadly amoeba.

A Girl Named Courtney

Courtney Nash was a teenager who knew what she wanted out of life — to be someone who helped make the world a better place. As a child she assisted handicapped kids through horse therapy, took care of her dying grandfather at the tender age of thirteen, and dreamed of one day being an OB/GYN physician. She always looked out for others and kept one eye on her future, while also enjoying just being a youngster.

On any given day, she could be found playing volleyball or softball, fishing with friends, hunting during open seasons, and simply being a part of the great outdoors. And, like so many teenage girls, Courtney was the self-professed biggest Taylor Swift fan on earth.

Whether she was playing sports or hanging out at family game night, planning her future or studying hard for an exam, going out with friends or for a horse ride through the wooded trails, Courtney lived her life to the fullest, and with a devout devotion to God.

"If something does not go as planned," she once wrote for a high school essay about her plan for the future, "I will be fine with that because I know God has a path leading me to my future."

For Courtney, her dream included attending the University of Florida, earning her nursing degree and getting a job in the medical field, marrying her high school sweetheart and having three children, and living on a farm with lots of animals.

The dream came to an end on August 3rd, 2011.

Along with her friends and family, Courtney went swimming in the St. Johns River in Titusville, Florida, enjoying the warm weather and jumping off the dock into the same river that had been home to many adventures in the past. The idea of potential dangers waiting in the water was not a thought on anyone's mind, for what is there to fear in a river that has always been a place of entertainment and family fun?

The day Courtney came in contact with the amoeba that would soon take her life, she swam alongside her brother, Ethan Nash, the siblings enjoying the day without a care to the world as they had done so many times before. But only one sibling would walk away from this day unscathed, for August 3rd would be the last outing Ethan enjoyed with his little sister, who looked up to her big brother as her best friend, her hero, as she so named him in an essay for school.

"A hero is a trustworthy and honest friend. He never gives up, and he fights for what he wants. My brother fits in all of these categories," she wrote of her brother, Ethan. Courtney appreciated how protective he was when she went out on dates, looked to him for help with homework and tests, and was encouraged to succeed with his support. "I have gone to every [wrestling] match he has ever had," her words continue. "He even did the same for me in volleyball."

This close sibling bond made the Nash family that much stronger — and made Courtney's loss that much harder for both parent and sibling. As much as they depended on one another, looked to one another for guidance and support, there was nothing any of them could do to stop one of their own from being taken away from them.

Only five days after that innocent swim in the water, Courtney began experiencing one of the first symptoms of the amoeba: headaches. Eight days later, she passed away after being diagnosed with a disease that, until the diagnosis had been made, no one in her family even knew was a threat.

The diagnosis — the amoeba known as *Naegleria fowleri* — had entered Courtney's central nervous system and made its way into her brain, which led to an infection known as Primary Amoebic Meningoencephalitis (hereafter referred to as PAM), an acute and necrotizing infection that leads to death in approximately seven days. Together, this amoeba, and resulting infection, claimed her life.

Chapter 2

An Overview

Naegleria fowleri, commonly known as the "brain-eating amoeba," is a single-celled organism found in bodies of freshwater that reach or exceed 80 degrees Fahrenheit (though it is important to note that some cases have been recorded in temperatures below 80 degrees). These areas include rivers, lakes, hot springs, mud bogs, and poorly maintained residential or commercial swimming pools.[1] The amoeba can also be found in moist soil, sediment, and even rotting vegetation, as the amoeba can be aerosolized in both natural and artificially heated waters or power plant discharge. Other water systems may include contaminated well water — and the soil the water feeds into — as well as sources one may least expect, such as old, homemade, or heated contact lens solution or even isolated in dust.[2]

It is simplest to say, perhaps, that this amoeba is capable of growing and surviving in the widest range of freshwater and brackish areas as possible, in areas we most and least expect.

Naegleria fowleri thrives in higher water temperatures, often as high as 115 degrees Fahrenheit, in all of these systems. The amoeba is able to survive in such diverse areas by feeding on microbes and bacteria within the environment. Quite literally, it is a bacteria-consuming organism. Consequently, it is not able to thrive in saltwater systems, such as the ocean, and instead is commonly found in southern states of the US as well as areas with many freshwater lakes, rivers, or tributaries.

The amoeba was named after Australian pathologist Dr. Malcolm Fowler, who, in the 1960s, was the first to isolate the amoeba in brain tissue samples from patients who died of PAM. Since identifying the amoeba and subsequently giving it its name, *Naegleria fowleri* is the only strand of *Naegleria* known to infect humans. While it typically feeds on bacteria within the water, in rare cases it has been known to infect people by way of the nose (particularly in instances where water is forced through the nasal passage, such as leaping into the water off a dock or falling after water skiing).

A higher percentage of reported infections (more than half) have taken place in Florida and Texas, though others have been documented as far north as Minnesota and as far away as Australia, among other foreign nations. The majority of those infected (more than one hundred cases) have been children under the age of eighteen, and more than 75% have been male. There are theories as to why one age group or gender may be more susceptible to coming in contact with the amoeba, which will be discussed later in this book, but thus far no concrete data has been provided to prove one theory correct above all others.

Nearly all victims, regardless of age or gender, were reported to have engaged in water-related activities, primarily during summer months of July and August that may have resulted in water going up the nose, such as diving, skiing, or simply swimming underwater. Other infections were traced back to the use of nasal rinses. A related, yet not as closely studied, aside — infections have also been reported in cattle and a tapir, though experimental research has not yet been fully conducted to conclude studies on this matter. For the purposes of this book, we will focus primarily on human contact, though it is important to note that PAM may be an infection capable of jumping species.

Naegleria fowleri is contracted via the nose rather than by drinking contaminated water. Once the amoeba enters the nose, it attaches to the mucosa and migrates to the brain via nerve endings in the olfactory nerve, working its way through the mucus and cell junctions without destroying cells in the process (which may be a factor in why it takes symptoms so long to appear, as the body doesn't detect anything actually wrong).[10]

The amoeba's ability to pass through without destroying cells is actually somewhat exceptional, as it allows the organism to not only protects itself, but also ensures that the nerve endings act as a sort of carrier for the amoeba, directing it to the central nervous system and directly into the brain. Potentially the amoeba could enter any area with a large blood supply, which would allow it to spread to other organs, but it is typically confined to the brain. Finally, the amoeba travels through the cribriform plate in the skull (the plate that separates the brain from the sinuses), here leaving behind extensive tissue damage along the path of invasion.

One primary attribute of amoebas is that they are "eating" organisms. They consume anything in their path — algae, other bacterial-type materials, organ tissues. The brain is a rich source of food, and so once it enters, the amoeba begins to dissolve and essentially destroy tissue and cells by excreting specific proteins and enzymes. The amoeba then feeds off this debris, which it needs in order to survive within the body, resulting in tissue loss and the severe brain infection known as PAM. [4]

PAM

PAM is considered a disease of the central nervous system, often taking on the same flu-like symptoms as meningitis in its early stages, which can result in misdiagnosis upon initial testing. Once the amoeba is detected, victims may have symptoms that include high fevers, respiratory problems, rashes, intracranial pressure, seizures, hallucinations, and loss of brain activity. Research still needs to be done on PAM and *Naegleria fowleri* to help medical professionals identify the infection in its early stages, as life expectancy after contracting the amoeba is often less than fourteen days.

Throughout the years, *Naegleria fowleri* has shown resistance to changes in habitat in its fully matured stages, surviving well in various locations provided there are warm bodies of fresh or brackish water present. These conditions allow the amoeba to survive regardless of environmental changes, and because it is a free-living organism, there is no way of controlling potential amoebic populations in rivers, lakes, springs, or pools.

The amoeba's ability to thrive within these water systems allows it to reside in places not many would suspect. While *Naegleria fowleri* is mostly common in southern US states where water temperatures tend to be warmer, it can just as easily be found in northern areas, surviving in the sediment in wait of warmer seasons. However, many people are not aware of just how easily the amoeba *can* exist in colder climates, which can lead to a false sense of security.

For example, in Minnesota there have been five deaths in five years related to PAM. Minnesota, the Land of 10,000 Lakes (though that number is likely more accurate as 25,000 to 30,000), may be one of the last places anyone would suspect of harboring such a deadly amoeba that lives in warm, if not hot, waters — and that fact may just as easily lead to the belief of "this can't happen to me" and proper precautions not being taken.

Of course, this is not to say that every river, every lake, every body of freshwater contains *Naegleria fowleri*. "We don't want to scare people or be an alarmist," states Heidi LaMeyer, chair of Swim Above Water, a Minnesota foundation dedicated to raising awareness for PAM while honoring the victims of this deadly infection. "But we hope that people will take precautions when they are in those bodies of water."

Not All Victims Are Children

In most cases, the amoeba is contracted during warm summer months while engaging in sports-related water activities such as wakeboarding — and it doesn't always affect children, though they are the primary victims. Waylon Abel of Indiana passed away at age 30 after living with the infection for nearly a month. Similarly, on July 31st, 1991, Mark Kinkade of Texas passed away at the age of 26. Mark was young and full of life, his widow Deanne remembers. He loved swimming, fishing, and water skiing, and it was during those events that he came in contact with *Naegleria fowleri* after swimming in Lake Granbury.

Mark and his family went to the popular lake

every weekend, staying at a friend's lake house while enjoying long days on the water. On this particular day, Mark, Deanne, and their ten-month-old son went out on a boat, alternating between skiing and swimming. Though it's not clear what part of the lake, or during which part of the weekend, Mark contracted the amoeba, his death can be traced back to those two days.

Seven days later, on a Saturday, he began showing signs of illness, beginning with a headache.

"I'll never forget," Deanne states of that moment when she first realized her late husband was sick. "I came home from running errands and was surprised he wasn't out mowing the yard. I walked in the front door … and saw him sitting on the couch, knees to his chest, face in his hands. I asked what's wrong, and he said his head was killing him." The pain was so bad, she recalls, that he had fingernail marks in his forehead from gripping his head so tightly.

Deanne brought Mark to the hospital the next day, where doctors gave him pain meds that made him sleep, and eventually put him in a drug-induced coma. During the next three days, doctors would have no idea what to look for. One hospital doctor gave a diagnosis of meningitis and encephalitis, and even stated that it was caused by a virus, and *not* a parasite, which ended up being the cause. Though doctors weren't exactly sure what was causing his illness, they did suspect it to be contagious, and placed Mark in isolation for the remainder of his time. For the next few days, Deanna would only be allowed to see her husband a couple times, always wearing a mask

"It was over before I knew it. I don't think he even knew he was in the hospital," she states, unsure even decades later if he was aware of his surroundings.

Mark ended up passing away just a few days after being admitted, but medical officials didn't know until months after what he actually died from. An autopsy was ordered immediately by the county health department, but since that didn't show a cause of death, brain slides were sent to Southwestern Medical in Dallas, where pathologists found *Naegleria fowleri* in the brain tissue.

This was the first time the amoeba had ever been talked about since Mark fell ill. Until this point, Deanne had never even heard of the parasite, and as the case was back in the 90s, nor was it really talked about in the media. Deanne spoke with a family doctor after getting the official diagnosis, and this doctor confirmed his own fears by saying he would never allow his family to swim in a freshwater lake in Texas.

Such is a lesson Deanna carries with her to this day, not having returned to freshwater and holding her nose even in saltwater, all despite the fact that her area of Texas has strict water treatment regulations. She fears for her son, now in his early twenties, who still ventures into rivers and lakes to ski with friends. "He knows the story and what happened [with his father]," Deanne speaks of her son. "I keep him educated and aware, but he didn't live through this like I did."

Deanna has had to accept that her son still enjoys swimming in the water. As he has said to her, "Lighting doesn't strike twice in the same place. This isn't going to happen to you again." But, as a mother, that worry never goes away and the fear of losing another loved one is always there. Deanne keeps up with stories of *Naegleria fowleri* victims, calling each case "frighteningly similar" to what she went through.

The memories are still fresh, perhaps lending to her fears for her son. "I wish we had gone [to the hospital] sooner," she recounts of that time more than twenty years ago. "I could have saved him some pain, probably, but I don't think I could have saved him."

And so now she does what she can to save her son, her friends, those around her, just like the loved ones of victims across the US do each and every day.

These are just two examples of adults contracting the amoeba among such a higher statistic of children passing away due to the infection. *Naegleria fowleri*, though, isn't picky in who it infects. It survives wherever it can and takes advantage of poorly maintained or overheated water systems. Because the amoeba is a free-living, widely-distributed organism, health management is not always prioritized when it comes to water samples and testing, and is often worse at the community level. Many microbiology concerns focus on water treatment to A) produce potable drinking water and B) mitigate waste discharge into natural water systems.[2]

Naegleria fowleri is not a danger in terms of drinking water, which means it may thrive in systems used for consumption. But when that potentially infected drinking water is used for other purposes, such as bathing or for nasal rinses, prior water treatments are for naught, as this amoeba has proven throughout its history to strike even in the most random and innocent of times in victims of all ages.

Chapter 3

A Change in Climate

We must take this history into consideration, Dr. Sandra Gompf states, especially when applying it to current prevention and treatment applications. The amoeba has already proven to show resistance even in colder temperatures, so what happens when the earth's climate shifts and allows for a more hospitable environment overall?

Studies have already shown that global warming is a very real threat, and that the climate is getting hotter. Across the United States, record-high summer temperatures are being recorded, which means a rise in water temperatures — which means more people are swimming in potentially dangerous water systems. It also poses the danger of areas that may not have previously seen higher temperatures suddenly being exposed to the amoebic threat.

Though the disease is currently classified as rare, which means it is one that has impacted less than 250,000 people (such as Tourette Syndrome and Lou Gehrig's disease), the fact that the climate shift is allowing for a more amoeba-friendly environment is a threat to people worldwide.

This theory is further strengthened by the fact that other diseases are already appearing in higher numbers alongside the global climate shift, while others are spreading for reasons not related to temperature changes but simply an ever-growing human population and related factors.

For example, malaria is appearing in higher numbers than previously recorded, while viral illnesses normally seen in tropical parts of the world are being recorded in other areas. Dengue fever is one of these diseases that is moving up in latitude from South America, spreading and expanding into northern territory, such as the Florida Keys. Both are mosquito-borne diseases, which shows that mosquitos as well as diseases are moving north, but since the amoeba is a free-living organism, it has the potential to live wherever it desires and potentially infect anyone.

But, if these viruses can spread, either due to human population, travel, or climate change, then so too must we consider the spread of *Naegleria fowleri* cases. The difference between the amoeba and the prior-mentioned diseases is that the amoeba can potentially live in any warm water system and may already exist across the world, only needing a simple shift in climate to accelerate its growth. What was once a rare disease may become a more common cause for concern in the future, and we will need its historic data to fuel the push for better prevention and treatment methods.

It can be hypothesized, though, that the increase in *Naegleria fowleri* cases is due to "changes in environmental conditions, thermal pollution of water from industry, and the development of industrialized areas with nuclear power plants and cooling towers that allow for concomitant greater growth" of the amoeba and the bacterial food sources it thrives upon.[10] In other words, *we*, humans, are partially responsible for the amoeba's presence by allowing our waterways to be so polluted, contributing to climate shifts, and not recognizing the fact that PAM is a very real threat that *should* be declared as a Nationally Notifiable Disease (though, many people do recognize this threat, but are the average man and woman, not governmental decision makers).

Of course, at this moment the above is simply theory. There is no conclusive data to show that global warming will lead to more cases of PAM, but there are factors that lend themselves to building this research. With a rise in temperatures comes more opportunities for people to swim in shallower freshwater, which means increased exposure to *Naegleria fowleri*. This is not to say they are automatically going to contract the amoeba, but simply that the risk is there.

"We can't make a logical guess that there will be a rise [of infections]," Dr. Mobeen Rathore says, as researchers can't say for certain if a potential rise in PAM cases is a direct effect off climate change, or an indirect effect.

Environmental specialist Bob Walton agrees. "Any 'beasts,'" as he nicknames the wide range of single-cell organisms, "known to be most active when moisture and water temperatures are highest will profit." Warmer annual climates, plus adequate water temperatures and ponds rich in organic debris, will certainly benefit free-living amoeba. And, of course, more people will be swimming.

The theory is credible, he states, but the best evidence will take years of warming to sort out actual and specific trends, such as a rise in *Naegleria fowleri* cases.

An Accidental Host

"[The infection] is still so rare," Walton explains. "While the amoeba is common, the infection in humans is rare, and accidental. Therefore, it is hard to get research funding or a significant sample size to study such rare occurrence outcomes."

There are options for studying the amoeba. Currently, *in vitro* studies (in a test tube or lab) have been performed via cell cultures. In tests, the amoeba trophozoites destroyed nerve cells along with other cell types. This helps researchers better understand the organism, but ultimately *in vivo* studies (in a mammalian organism) are needed to better see what treatments will be most effective, especially considering the amoeba "has developed mechanisms to evade the host immune system," as evidenced by its ability to pass through the nose without destroying cell junctions, [10]

Naegleria fowleri isn't an opportunist organism, Walton continues, as it typically presents itself in otherwise healthy individuals. An opportunistic organism would seek out the weak or ill, since the immune system is already compromised, but this amoeba has no standard for its host. Healthy or unhealthy alike are victims, provided the climate is right.

Regardless of climate, there are probably billions of encounters between people and the amoeba each year in Florida alone. "If you've been swimming in the right waters, you've probably been exposed," he warns, the 'right waters' being any body of freshwater that has reached 80 degrees Fahrenheit — though this isn't necessary a hard and fast temperature to always go by.

In fact, there is a high probability that those who do come in contact with the amoeba, come in contact with *several* amoeba at once. It is a slim chance that those who passed away from *Naegleria fowleri* died of just one amoeba. Instead, they likely contracted several amoebas, which then reproduced rapidly within the body as they made their way to the brain.

In this way, humans are very much an accidental host. This free-living amoeba feeds on bacteria and doesn't seek out human contact. But, once it does enter the body, it does what it knows best — destroy tissue for consumption.

So, if we are accidental hosts capable of contracting multiple amoebas at once, we must take proper precautions, especially in regards to potential climate change. While we don't know everything about the amoeba, we do know facts about its environment.

Naegleria fowleri feeds on organic material, and organic debris is what settles into pond sediment, where bacterial decomposition consumes oxygen and renders larger organic materials (such as fish and plants) into microscopic material. Altogether, this makes for a perfect amoebic home, and when that home is disturbed, the amoeba has a higher chance of invading human bodies.

To counteract this, continually flowing water is the best chance of avoiding contact. Fountains in ponds, aeration systems in lakes, anything that produces bubbles and water flow will oxygenate the water and disrupt favorable conditions for the amoeba. In this way, whether water temperatures continue to rise or not, we can do our part to ensure those systems are as safe as possible.

Chapter 4

The History of PAM

PAM infections have been reported as early as 1961 in Australia, with a second fatal infection being reported in 1962 in the United States, in Florida. Upon further investigations into the amoeba, along with tissue sample research, it is believed that PAM may have been the cause of death for cases as far back as 1937 in Virginia.[3]

Since the 1960s, as far back as can accurately be traced, there have been nearly 130 cases of PAM infections, all fatal except for three. Many of these infections were reported in the southern United States, including Florida, Georgia, Louisiana, Mississippi, Texas, and South Carolina, with additional cases reported in Arkansas, Arizona, California, Indiana, Minnesota, Missouri, North Carolina, New Mexico, Nevada, and Virginia. Nearly half of these cases occurred in Florida and Texas, with the majority of those affected being men and children (though there is no research to support age or gender being a primary factor in contracting the amoeba, this statistic may be a reflection of the population that typically engages in water sports).[4]

Sadly, it doesn't have to even be extreme water sports, which one may expect, as the likelihood of water being forced up the nose is much more common. Sometimes, just swimming and horseplaying with friends in a local watering hole is all it takes for one amoeba to take its toll.

In August of 2012, nine-year-old Jack Ariola of Forest Lake, Minnesota, went swimming with his family in a Stillwater lake. Jack was an active little boy who loved building things from scratch and playing sports, especially wrestling, baseball, and, most of all, hockey — and was even photographed in the Stanley Cup in 2003 to commemorate his passion for the sport. As a little boy, Jack even promised his mother that one day he would build a house for her, showing that even at a young age, he was a kind-hearted soul who loved his family. And, like many other victims of *Naegleria fowleri*, he loved being in the water.

There were many children and families innocently swimming that day, unsuspecting of any potential threat lurking in the water. With his own family, Jack enjoyed a day of fun like he'd done so many times before. But this time, that day came with consequences no one could predict. Only a couple days after swimming, Jack began experiencing headaches and dizziness, which later progressed to vomiting.

Suspecting a severe case of the flu, Jack's mom brought him to the emergency room. Due to his combination of symptoms, doctors initially suspected meningitis. When his symptoms advanced, the medical staff began questioning Jack's parents about his activities the past few days, including whether or not he had been swimming. This was an interesting question asked in order to diagnose the boy, as oftentimes whether a patient was swimming or not is never a concern until all other options have been exhausted, such as in Courtney Nash's case. For Jack, it was the question needed to help determine that he had contracted the deadly amoeba, which resulted in a fatal brain infection (PAM), following exposure to *Naegleria fowleri*.

Despite being in a far northern state, during a time when water temperatures weren't necessarily at a concerningly high level, the amoeba still thrived. Even as a single-cell amoeba, or perhaps because of that, *Naegleria fowleri* was able to adapt and survive in the sediment. And because there is no cure for the single-cell organism that causes PAM, Jack passed away on August 6th, 2012.

Jack's death gives strength to the question asked by so many families of victims — why do we not know more about this amoeba, and why are people not made aware of this danger before allowing their children, or themselves, into the water? For so many of Courtney's family, including the one putting these words to paper, the amoeba and its devastating effects were never spoken about through an entire lifetime living in Florida. For so many of Jack's family, the question of *why* is one they fight to have answered each and every day.

Signs are typically posted at popular swimming locations, Jack's father, Jim, recalls. Small 8x10 posters warning swimmers to swim at their own risk, but these risks often aren't actually outlined. This simple fact poses the question, are these lakes or rivers (or any bodies of water) actually tested? If so, what are they tested for? And, if they are tested for any sort of potentially hazardous materials or ecological dangers, why are these specific risks not posted?

Since Courtney's passing, county officials did agree to put up signs at different parks and boat ramps along the St. Johns River around town, but this is not a common occurrence. In fact, many people are not warned about fatal infections like PAM. Lakes and rivers are rarely shut down in the wake of a death by amoeba, as evidenced by Mark

Kinkade's passing. No alerts were sent out, no temporary shutdown for testing, because, as Deanne remembers, "Everyone said it's so rare compared to the number of people swimming that it's not going to happen again."

A Question of Rarity

For friends and loved ones of *Naegleria fowleri* victims, the very name, let alone the dangers, are never even heard of until the amoeba has already taken its toll. Some believe that more about the amoeba is known than is let on — and that the truth isn't told because it would affect numbers at popular swimming locations, whether local lakes or famous water parks in tourist towns. People would be scared away, and, along with their fear, so too would the profits disappear.

With an amoeba that has its traces as far back as the sixties and with more than one hundred known victims, one may question why more about the parasite and resulting infection are not talked about. Is PAM by way of *Naegleria fowleri* as unusual as people are led to believe?

"It is as rare as people think it is," Dr. Rathore agrees after considering past research, chances of misdiagnosis, and incorrect death certificates. "It's a rare thing no matter how you cut it. You need to have the right circumstance, the right ecology, so that everyone has had that exposure." In other words, there are several factors at play in actually coming in contact with *Naegleria fowleri*, and for as many people who were exposed, there are that many who were not.

There may be several, or any unknown number,

of people who died of PAM that research doesn't know about, he acknowledges, but right now we aren't seeing hundreds of cases a year directly related to PAM. If we were, then there would be more research.

Along with this, we may need to consider the very definition of "rare" itself — 250,000 seems to be a high number, but when we consider the number of death-by encephalitis cases each year that have no known causes (a shocking 86.2% from the years 1989-1998)[15], is it so hard to believe that cases of *Naegleria fowleri* aren't to blame? And if there is cause to suspect more cases are out there, but not being diagnosed, is PAM really that uncommon?

"It's not that difficult to diagnose," Dr. Rathore offers, "but you can only diagnose it if you suspect it." So, while there may be opposing viewpoints on how sporadic or common *Naegleria fowleri* might be, there is the agreement that regardless of its status, people need to know about it.

It is important to note that the amoeba itself is *not* rare. It is found across the nation, and even the world, in warm waters. When medical teams or scientists discuss the concept of rarity, it is often in regards to the infection that results from contracting *Naegleria fowleri*. The rarity of PAM, however, is still up for investigation and research. And with these investigations comes the question, how do we make sure the public knows?

"We can't keep it in," Jim Ariola states firmly. "We have to tell their stories." Telling the stories of each victim is the only way to bring awareness to the public, while remembering the ones lost. We must give a face to the disease, a face of children whose lives were taken too soon, a face of parents who will always grieve for the little ones left behind.

"Still, to the day, it's unbelievable," Jim continues. Jack's little brother talks about him all the time, his father expresses with sadness, but how does a parent explain to a younger child what happened to his big brother? "I see his little buddies every day, the kids Jack played hockey with. I see the same people all the time. Everyone's been so supportive," he recalls, "but they ask, when is Jim gonna go back to his old self? It doesn't happen that way, you don't just kick this and get over it."

For Jim, there is no "getting over" his son's death or going back to the person he used to be, and so he does what he can to ensure other parents don't go through this same trauma — by getting involved with organizations devoted to raising awareness, and being the voice people need about a silent killer that has flown under the radar for far too long. To accomplish this, Jim has joined the Swim Above Water organization, taking over as vice president.

As a member and director of Swim Above Water, Jim seeks to educate everyone about the potential dangers lurking in warm water systems. He gives out free nose plugs, helps host awareness events, and tells everyone to be careful when swimming. Even though Jim says he, personally, will never go in a lake again, he also assures, "I don't want people to be afraid, just safe."

Little Jack is with him in spirit, helping Jim and Swim Above Water in their goal to educate the public and bring awareness to the dangers of *Naegleria fowleri*. "Life is not guaranteed," Jim seeks to remind everyone. "Give your loved ones hugs and kisses, because you never know, it could be your last time."

With his foundation, his memories, and his mission, Jim helps pave the path to a safer and more promising future, fighting to find a way to test for, diagnose, and treat this deadly amoeba that has taken so many lives in the past.

Chapter 5

Testing Waters

Since 1962, more than one hundred and twenty-five PAM infections in the U.S. have been associated with drinking water systems used for swimming, bathing, and cleaning out nasal passages, as well as illnesses caused by swimming in warm freshwater systems, such as lakes and rivers. Australia has similarly seen PAM infections in the 1960s. In Australia, infections were linked to drinking water being piped overland for long distances, which resulted in the water being heated to a temperature that allowed for the amoebic population to thrive.

While homeowners may not suspect pipes of being capable of harboring life-threatening parasites, the plumbing and irrigation process may lend itself to providing favorable conditions for an amoeba that will later enter a human bloodstream. Because of this fact, some water systems throughout the Australian country are regularly monitored for *Naegleria fowleri*.

This protocol was enacted by the Government of Western Australia's Department of Health and Public Health and Clinical Services. Its goal was to accurately identify the amoeba in reticulated water supplies, detecting *Naegleria fowleri* in drinking water in order to take fast corrective measures. Any detection of the amoeba in any water system is an indication to officials that preventative measures failed.

Studies have shown that, in addition to growing in waters of high temperatures, *Naegleria fowleri* is able to enter a water supply system at its source or in cracks of tanks or reservoirs. The amoeba can even thrive in pipework and tanks in an environment that is hospitable to its lifecycle.

The Amoeba Response Protocol, enacted by the government, set out to catch the amoeba where it slipped through the cracks, chlorinating and disinfecting water supply systems consistently. Routine monitoring, especially during months when the water temperature is expected to rise, is especially important, and when traces of the amoeba are found, advisories are sent out via email or facsimile by the Department of Health.

These notices, in short, advise medical facilities and professionals of the amoeba in drinking water systems within the specific communities. These notices are sent as a courtesy, even if actual contact with the amoeba isn't likely. Similarly, the Department of Health sends a notice to the community itself, warning the members of the weather and elevated water temperatures that may lead to a rare, albeit nearly always fatal infection, known as amoebic meningitis.

With this notice comes advice to the community members on proper care and prevention, such as not allowing water to go up the nose, not jumping or diving into the water, not allowing children to play in the water unsupervised, and making sure pools or bathtubs are properly cleaned regularly. These are just general precautions the Australian government offers, along with tips on safety and awareness.[9]

This may seem like extreme precautions to some, considering the disease's 'rare' classification, but it is notable that there have been only four recorded cases of amoebic meningitis in Western Australia up to 1985, when it became an official notifiable disease — and no additional cases since.

This is proof that when adequate prevention and public education measures are taken, the risk of contracting *Naegleria fowleri* drastically reduces. So why, then, does the US not list PAM as a Nationally Notifiable Disease and put the same measures into effect?

"It's simply not feasible," Dr. Rathore informs. Australia focuses on treating water used for residential and potentially commercial use, "which isn't typically where the amoeba is found, as it shouldn't even be there." But, while it shouldn't be there, proof can be shown that the amoeba is there regardless. The fact remains that *most* cases of amoeba contact take place in a river or lake, and treating it would require chlorinating freshwater systems across the United States. This would be an enormous undertaking, and at this time, that undertaking may actually be unrealistic.

Considering the number of freshwater systems in the nation, it is not surprising then that there simply isn't an effective way to treat our waters. As a result, it is up to individual cities to put preventative measures into effect. Several recent news releases describe legislation that has recently been enacted in two Louisiana parishes to improve water testing and cleaning systems, which proves that some efforts are being made within various locations. However, this legislation is restricted to those areas and not the nation as a whole. This may mean that the number of infections per year will rise, particularly if summer temperatures continue to rise.

The yearly statistics have been attributed to specific heat waves, which increases water temperatures. This number may range from zero to eight victims per year, though the average has not increased since 1978, when the CDC established its Free-Living Ameba (FLA) Laboratory and PAM registry.[4]

A Silent Killer

Regardless of this average, regardless of whether or not the water is treated, the fact remains that *Naegleria fowleri* is a silent killer, attacking during a single, innocent swim amongst friends. For Courtney Nash, just one week separated a promising future as a nurse from a family's decision to let her pass on.

After her August 3rd swim in the St. Johns River, Courtney went about her normal teenage life with no idea of the parasite making its way through her bloodstream and up to her brain. Within the next few days, Courtney registered for the upcoming school year, obtained her driver's license, and reconfirmed the organ donor status — she had been a donor since the age of fourteen, a status that was extremely important to her.

It was that following Monday, August 8th, that Courtney began feeling ill. Her symptoms began with a headache, though not one that caused concern. After all, headaches are common enough, and not typically a sign of serious illness, suggesting stress or exhaustion rather than disease. But, within the next day, that headache escalated to an unbearable amount of pain made worse by constant vomiting (upwards of twenty times in one day).

At this point Courtney was still lucid, albeit aware of having some sort of severe illness. She even posted on her Facebook page, "103 fever +

major headache + throwing up = miserable" in one of the last posts she would ever make. It was then that her mother, P.J., took Courtney to the ER, and Courtney would make her last post of "is in the ER." Nearly forty replies would follow her post, wishing her well, engaging in conversation as much as possible with the teenager who, despite being so ill, took the time to thank everyone who offered a 'get well soon' message.

In the midst of her well wishes at the hospital, doctors took and tested samples of Courtney's blood while also giving her medicine for the pain: 500 milliliters of saline IV, electrolytes, and Rocephine (a standard antibiotic for meningitis). At the time, the ER doctor thought Courtney's illness to be a bad case of the flu, as the hospital had already seen six to seven patients showing similar symptoms recently, so she was sent home with the order to return if she did not improve.

In many, perhaps even most cases, returning home from the hospital is a good sign, a sign that one's illness is improving, or not of enough concern to warrant a stay in the ER. Typically patients can relax just a little, and wait for their bodies to heal. But for Courtney, her body only began to shut down.

After Courtney was discharged from the ER, her fever continued to rise throughout the night, spiking up to 104 degrees in just a few hours. A rash also broke out, which prompted Courtney's mother to contact her brother-in-law and District Fire Chief /Paramedic, Tom*, around 08:30 AM the next morning. At this point, everyone wore masks, worried that whatever Courtney had may have been contagious.

By the time Tom arrived, Courtney was suffering a high fever, a severe headache, neck pain and stiffness, and a rash across her back, though she was still responding to verbal stimuli. On August 10th, her family rushed her to Arnold Palmer Hospital in Orlando, Florida, where she was immediately admitted under the suspicion of meningitis.

This initial and tentative diagnosis was made because the symptoms Courtney showed were clinically similar to meningitis, which is serious and potentially fatal disease developed in response to bacteria or viruses, or even certain drugs. The disease gets its name due to it causing swelling of the meninges, which are protective membranes around the brain and spinal cord. This swelling is directly caused by an infection of the fluid around the brain and spinal cord, which is why a spinal tap is administered in order to check for traces of infection. Doctors at Arnold Palmer performed a spinal tap on Courtney, which produced cloudy spinal fluid that indicated some sort of infection.

Some cases of meningitis clear up on their own, while others require immediate medical attention. Because it is potentially life threatening, doctors must begin treatment as soon as possible when this disease is suspected.

Symptoms of PAM

Like with meningitis, symptoms for the *Naegleria fowleri* amoeba begin one to seven days after exposure, with death occurring within one to twelve days. These symptoms are placed into two stages based on severity:

Stage One:

Severe Frontal Headache
Nausea/Dizziness
Vomiting
Fever

Stage Two:

Stiff Neck
Seizures
Mental Changes
Hallucinations
Coma

Other Symptoms May Include:

Smell or Taste Abnormalities
Nasal Obstruction or Discharge
Photophobia
Lethargy
Loss of Balance

Naegleria fowleri is able to mimic the signs of meningitis because it is considered a free-living ameboflagellate — it can survive in both soil and water in extreme temperatures, and causes life-threatening infections in otherwise healthy humans by invading the brain and meninges. By entering via the nasal mucosa and olfactory nerve, it causes acute, fulminant hemorrhagic meningoencephalitis and the resulting PAM, which then mimics meningitis.[6]

Because these symptoms so closely mirror those of meningitis, the accuracy in properly diagnosing PAM is greatly lowered. It takes medical teams longer to rule out illnesses such as the flu or meningitis, but since PAM strikes so quickly, often times the proper diagnosis comes too late.[5]

Such was the case for fourteen-year-old Will Mathews of Louisiana. Will was an energetic teenager and older brother to three siblings. Always with a smile on his face, the boy was often seen playing sports (especially baseball, soccer, and football), hitting the trails on his dirt bike, and joining his family in celebration of their love for the Lord. Described by his family as a teenager who was always up for the next big challenge, Will's zeal for life was larger than most his age. And, in the summer months, that passion for living large and having adventures led to a day of fun wakeboarding on the waters of Louisiana.

Just one week after the outing, Will began showing the classic signs of PAM. As with Courtney, doctors initially misdiagnosed Will's illness, though in his case they suspected encephalitis rather than meningitis. He was placed on life support, and passed soon away on August 15th, 2010. "Not a day goes by that we don't miss our son greatly," says his family, "but [we] just have to put our complete trust in God and rest in the fact that we will all be together again."

Too many families have echoed these same sentiments, having lost their child, spouse, brother, sister, parent, to this deadly disease. And so many victims share one common factor outside of the cause of death — the symptoms did not automatically point to the amoeba, because the amoeba itself is, in a sense, a master of disguise. It takes on the form of meningitis, encephalitis, and even the common flu, evading doctors often until it is too late.

Chapter 6

The Spread

It is hard to estimate the risk of contracting *Naegleria fowleri* and the resulting PAM infection, considering the millions of people swimming in freshwater systems every year, the same rivers and lakes they may have been swimming in each year, or even on the same day, as a victim. Calculating that risk gets even harder when we consider how many cases may have been misdiagnosed as meningitis or encephalitis.

The seemingly random chances of being infected makes for extremely inconclusive data outside of the environmental factors that increase the risk itself. Currently, no method exists to accurately calculate the risk, nor is there a method to measure the number of amoebas that may be in the water, though traces of *Naegleria* itself can be inspected. However, some doctors, such as Dr. Sandra Gompf, have been working at being able to do just that — make the calculation by looking at both the probability and impact of the infection.

For example, the probability of contracting the amoeba is less than being struck by lightning or being attacked by a shark. The impact, however, is far deadlier. Swimming in a lake ten times every year for ten years in the summer puts you at a risk of *Naegleria fowleri* infection of 0.0005 %. This percentage is close to your odds of being struck by lightning or attacked. The difference, however, is that instead of 10% probability of death by lightening or 5% probability of death by alligator or shark, PAM kills nearly 100%.

But one thing these other accidents have in common — they are more easily prevented. We can see a shark or alligator, defend ourselves if necessary. We know the protective measures to take to avoid an attack, or to avoid being struck by lightning. We know not to run outside during a storm. We know not to swim in the ocean with an open wound, or to play at the shores of an alligator-infested lake. These things are tangible, easily understood in order to protect ourselves. But what about what we cannot see? These things are all but invisible to the naked eye, easily forgotten about and therefore far more deadly.

Just because we can't see the amoeba doesn't mean it isn't real, or that it is somehow less deadly in terms of survival. Just as we know not to run outside in bad weather, so too should we know to wear nose plugs when we swim or to properly treat water before a nasal rinse. *Naegleria fowleri* and PAM are very, very real, and can kill in less than three days' time.[14]

And yet, despite knowing all of this and being able to calculate a risk to some extent, the disease is still classified as rare. Because of the classification, public health officials in the US currently do not enforce swimming or water standards.

Some research has even shown bacteria and amoebic growth in residential pipes, though it's unclear if using this water for bathing, cooking, and drinking is a risk. Data thus far suggests the amoeba cannot be contracted by consumption (in terms of drinking), but the amoeba being in the water itself does still put people at risk should that water somehow be forced up the nose. Oftentimes, people aren't aware of what may be growing in pipes (considering it's not an area easily or often examined), so coming in contact with the amoeba at home or work is hardly a worry.

And, like so many things, if the risk is low, then the worry is pushed to the back of one's mind, along with awareness.

In the grand scheme of things, it is true that the risk of contracting *Naegleria fowleri* is very low, considering thirty-one infections were reported in the US between 2003 to 2012, while hundreds of millions of people swam in freshwater systems recreationally within that same decade. In comparison, nearly 40,000 people died via drowning between 1996 to 2005. This example is not given to trivialize the risk, nor to downplay the importance in taking proper precautions, but rather to highlight the presence of the amoeba and its reporting in line with scientific knowledge.

Perhaps because of the amoeba's random and rarer attacks in correlation with more common death statistics due to other, more well-known diseases or accidents, medical officials may not even suspect its presence until it is too late. For many victims, PAM was not considered a cause in the initial diagnosis, caught only after more extensive testing after time had already run out. But for others, all attempts at testing failed.

Such was the case with Hailee LaMeyer, an eleven-year-old girl with a lively smile who enjoyed sitting down with a good book as much as she did just being a kid with her friends. Passionate about the arts, Hailee loved to write stories and plays, listen to music, play the piano, and even get outside for a game of volleyball. And, like so many children her age, she enjoyed being in the water.

On a warm Monday afternoon in August 2008, Hailee was swimming with her friends at Fawn Lake, a small public lake just a block from the family's home in Minnesota. That day, upward of thirty kids, including her brother Justin, were also swimming, enjoying the sunshine and one another's company, but it was only Hailee who fell ill just two days later.

That Wednesday Hailee began feeling sick, starting with a headache, fatigue, and fever, and her condition worsened quickly enough that she was brought to Urgent Care not long after falling ill. At Urgent Care, she was initially diagnosed with pneumonia and given two different kinds of antibiotics. The next morning, her fever continued to rise and she began vomiting, so her parents brought her back to the emergency room. There she was treated with IV fluids while additional tests were performed, and was again sent later sent home after doctors thought her condition was improving.

However, just an hour later she began vomiting again and displayed neurological symptoms (slurred and garbled speech, unable to walk, dizzy, and seizure activity). Heidi and her husband, Chad, rushed Hailee back to the emergency room, where doctors finally did a spinal tap and gave her a diagnosis of either meningitis or encephalitis (probably encephalitis). No further testing was done for *Naegleria fowleri* or any other infection during her hospital stay. Instead, she was then transferred from their local hospital to St. Paul Children's Hospital, where she was admitted to the PICU.

Her condition continued to deteriorate and she was put on life support, but just twelve hours after arriving in the PICU, Hailee's brain had died. From there, her parents made their most difficult and heart-breaking decision, to remove their daughter from life support.

The medical examiner declined an autopsy, and so Hailee's death certificate labeled her official cause of death as meningoencephalitis (a combination of encephalitis and meningitis), though this was a best-guess diagnosis. The family allowed doctors to take additional blood and spinal fluid samples for further testing, all of which were sent to the Minnesota Department of Health Unexplained Deaths Project for review. "No one even asked us the whole time if she had even been swimming, and not once did we think to tell [the doctor]," Heidi LaMeyer, Hailee's mother, says, recalling the mere five days it took for the amoeba to take its toll. "They asked if she had been around animals … if she'd been out of the country, but not swimming."

At the time, the LaMeyers were forced to accept the fact that the cause of death wasn't 100% positive, but they wouldn't accept not knowing. In 2008, not much press or awareness was given about *Naegleria fowleri*, and so research wasn't as readily available as it is now. For this reason, Hailee's parents kept up their search for as long as it took to get an answer, scouring as many medical sources as possible for some kind of clue.

It wasn't until two years later that the diagnosis was even able to be determined, as posthumous testing held no conclusive results. Original samples had been mishandled by the laboratory doing the test on her spinal fluid, as it had been left out at room temperature then frozen, and later thawed and frozen again during the testing process — all of which would have resulted in the amoeba bursting, unable to be detected. From here, there were no more samples to test.

This didn't stop the LaMeyers, though, as they continued searching for an answer based on Hailee's symptoms and activities prior to falling ill.

Reopening the Case

In 2010, Annie*, also of Minnesota, passed away from a confirmed case of PAM. Because of the similarities between Annie's case and Hailee's, health officials were able to return to Hailee's case. Ultimately, doctors were able to determine extreme likelihood that Hailee died of a PAM infection, despite the cause of death listed on her death certificate.

Hailee's case was able to be reopened because of her parents' diligence in determining the cause of death. The day the LaMeyers heard about Annie, they knew what avenue they could now pursue for their daughter. In the two years since Hailee's death, more information about the amoeba had surfaced, and so while a lack of knowledge or understanding of the infection within the medical community may have resulted in Hailee's original misdiagnosis, the LaMeyers could now pursue a more exact diagnosis. And they pursued it relentlessly, determined to discover the cause of their daughter's death.

"I wish we would have known what happened to Hailee right away, because two years of not knowing was really tough," Heidi says, looking back on those years. "That was two years of time we could have been saying what happened and educating others." Not willing to give up, Heidi and her husband did everything in their power to find out what took their daughter's life, being told by the health department that, though her case is officially undocumented with the MDH, they are 99% certain the amoeba was to blame.

This diagnosis encouraged the LaMeyers to continue with their cause of education while learning as much as possible about the amoeba. "You need to be an advocate and test for it," Heidi continues, speaking of her message to all parents to be aware of the dangers of swimming in warm waters. "Know the signs and the symptoms. Ask to be tested when the symptoms are present and freshwater exposure has happened. Especially now with a potential effective drug for treatment. Doctors also need to be educated on what to look for and test early for diagnosis for this rapid infection."

Part of knowing the signs and symptoms is knowing when to be cautious. For many, this means knowing and accepting that *Naegleria fowleri* can exist even in colder climates. Hailee and Annie, as well as Jack Ariola, lived in cold northern states where the water is cooler through June, as have several other children who contracted the amoeba and resulting infection. This knowledge is part of Swim Above Water's goal — to raise awareness that even in northern states *Naegleria fowleri* can thrive due to the amount of freshwater bodies. Heidi, who joined Swim Above Water about six months after its inception and now serves as chair to the organization, often works closely with other awareness organizations, such as the Courtney Nash Amoeba Awareness Foundation, to spread the message nationwide.

"We are all working together for the same goal," she reflects, "and we all have similar messages." While the infection is classified as rare, the risk of contracting the amoeba is extreme due to its volatile nature and nearly 100% death rate. There is also the question of how many undocumented cases there may be, given that the infection is often misclassified as meningitis or, in Hailee's case, encephalitis. In 2011, for example, a six-year-old boy contracted the amoeba after swimming in a freshwater Minnesota lake just three hours north of Hailee. After returning home to Iowa, he fell ill and later died only days later, on June 25th, of an unnamed brain infection. It wasn't until months after his death that officials were able to inform his family that PAM was the likely cause after exposure to *Naegleria fowleri*.

With record summertime temperatures in many part of the nation resulting in warmer freshwater bodies, the need to hear these messages is more important than ever. At Swim Above Water, as with all other organizations dedicated to awareness, the goal is the same: to educate the public on the risks of PAM while offering solutions on prevention and seeking better ways to diagnose and treat this devastating amoeba.

Chapter 7

Valiant Efforts

With the difficulty at hand in identifying the amoeba, and little public education to inform parents of what might be the cause of their child's illness, finally getting that diagnosis is another hard blow for families.

For Courtney's family, the spinal tap and resulting tests finally gave them that answer.

A short while after the spinal tap, her parents were brought into a conference room to meet with an infectious disease specialist*. The doctor, who wasn't initially assigned to Courtney's case, happened to be in the lab at the time the spinal tap sample came in. After personally studying the sample via microscope, he contacted the CDC along with amoeba experts around the nation. At the time of the conference, the specialist was already confirming his suspicion of *Naegleria fowleri* and was formulating a plan of attack with this team of experts. To this end he went above and beyond his role, taking on the case and scouring medical sources after hours to find the research that would save Courtney.

In the meantime, Courtney was transferred to the Pediatric Intensive Care Unit. The infection was later confirmed as PAM by PCR (Polymerase Chain Reaction) and culture, and upon receiving confirmation of the infection, doctors immediately began treatment. The primary focus was on restoring Courtney's airway via a breathing tube while lowering her body temperature. At the same time, the medial staff assigned to her care started a strict medication regiment ordered by the infectious disease doctor. Meanwhile, various specialists continued to research the amoeba for treatment options, considering the nearly 100% fatality rates of PAM coupled with Courtney's continually high fever.

Within the day, Courtney's symptoms transformed further. She was given a host of medications to treat cerebral edema (swelling of the brain), which later resulted in progressive lethargy. On August 11th she was intubated and administered additional medication via a lumbar spinal needle, and an external ventricular drain was placed to help alleviate intracranial pressure, which increased into the night. Eventually, a drain was placed in her head by a neurosurgeon to alleviate the pressure. This was the last step to be taken before experimental measures were put in place, as by this point the infectious disease specialist advised Courtney's family that he and his medical team had exhausted all known treatments of the infection proposed by the CDC.

At this point, many of the doctors and nurses, including the infectious disease specialist, began their own research, volunteering their own time by heading to the library instead of their homes after the end of their long shifts. None of the medical staff at Arnold Palmer were going to give up. Their goal — to help save Courtney, and uncover any knowledge that may have been overlooked in order to find the one treatment that would work. These solutions ranged from a fungicide to a more dangerous medication that may have damaged regular human tissue.

Despite the staff's selfless and valiant efforts,

by that following Saturday, doctors advised the Nash family that there was no brain activity.

The teenage girl once wrote, "I am guaranteed what my life is going to be in the future." This life included plans to attend med school and become part of the medical community that fought to save her life, and to later live in a two-story house in Tennessee with her high school sweetheart. Her future was to have three children (Landon Faith, Shyann Jean, and Jonathan David), and to own many animals, such as cows, horses, chickens, and ducks. That guarantee was taken away by a deadly brain-eating amoeba.

Courtney's death came just days after she posted a message on a family memorial board to her late grandfather. On July 30th, 2011, she wrote:

> *Hey grandpa,*
>
> *It's been 3 years and I can't believe it's been that long already, It's funny how time passes so quick! I miss you so much it's not even funny. There's not a day that goes by that I'm not thinking of you! I miss and love you so much! I can't wait to see you soon!*
>
> *I love you so much! P.S keep the signs coming! :)*
>
> *Love, Courtney!*

Within just a few weeks of this message, a message that eerily foreshadows the loss of a girl who clearly loved and cherished her family, Courtney joined her grandfather in a spiritual world she so closely believed in. Brain death was declared at 4:30 p.m. on August 13th, with an EEG and brain perfusion study already scheduled to conclude the doctors' findings.

Waylon's Case

These posthumous tests are bittersweet for many families, as they offer the closure so desperately needed, but may also take days, weeks, months, or even years to get the sought-after answer. Such a long length of time is primarily due to the fact that the symptoms are so misleading, so the autopsy is what gives the final diagnosis of PAM as a result of *Naegleria fowleri.* This is not to say that *only* the autopsy can provide results, as in some cases, others have been detected while the victim is still living via spinal fluid samples. But, for many, the diagnosis doesn't come right away.

Like Courtney, it took the family of Waylon Abel nearly a month to get that answer. Waylon was a family man, with three children and a doting fiancée, whom he loved to take fishing, swimming, and hunting. His brother remembers him as an all-around good guy, someone who was always smiling and made a friend out of everyone he met. More than that, he was always willing to help out complete strangers, and helped save a man's life after a car accident just weeks before his death.

This good soul, who so often made his friends and family laugh, was a great father, a loving brother, a spiritual man who often held Bible studies at his home. He enjoyed working with his hands, and could always be found enjoying Mother Nature while playing sports or hitting the trails.

An outdoorsman at heart — so much so that he refused to have a television in his living room so that this kids would be encouraged to play outside

— he could often be found in the water. It was during a single swim on July 15th, 2012, that he would come in contact with the amoeba. That summer day was extremely hot, and Waylon spent his time cooling off in a lake in Loogootee, Indiana, with his family. This particular lake is a popular spot for swimmers of all ages, and on the day Waylon contracted the amoeba, the shores were bustling with activity.

"I think about all the people in the water that day, and it's crazy," Sarah, Waylon's sister-in-law, says. "They talk about how rare [the amoeba] is, but it's everywhere in the water." On this day, only Waylon would come in contact with *Naegleria fowleri*. He and his family stayed off to the shallows of the lake where they could both fish and swim. Waylon alternated between fishing and teaching his kids how to swim, while also playing around as people often do in the water, doing cartwheels and flips into the lake. Waylon even commented on his own horseplay, jokingly stating at one point that it felt like the water had gone up into his brain.

This activity likely stirred up the sediment in a part of the lake that wasn't used as much, which in turn may have resulted in greater amoebic activity as Waylon swam. Regardless of what caused him to come in contact with the amoeba, just days later, Waylon began experiencing symptoms that included limb sensitivity. At first, he brushed off the symptoms as simply not feeling well. A little while later he began feeling like he had the flu, with a stiff neck and fever.

It was then, on August 5th, that Waylon went to the hospital. There, he was checked for a myriad of illnesses, including swine flu, Lyme disease, and West Nile virus, but was eventually sent home with a prescription for antibiotics. When his condition worsened only twelve hours later, with a high fever and vomiting along with nearly numb limbs, he returned to the ER and was placed on a ventilator, then airlifted to St. Mary's Medical Center in Evansville, Indiana.

While at the hospital a second time, Waylon held on as long as possible, but his body began to shut down. He went into cardiac arrest twice, and was resuscitated twice, during which time he was placed in an induced coma and his lungs began filling with liquid. Doctors advised his family that, at this point, the only option was to take him off life support, as the only thing keeping him alive was the adrenaline being pumped into his body.

Just a few hours later he passed away, on August 7th, 2012, five days before his thirty-first birthday. His death marked Indiana's first recorded death due to *Naegleria fowleri*.

At this point, however, doctors didn't have an answer as to his sudden illness. In fact, *Naegleria fowleri* was never even a suspect. An autopsy was performed, while at the same time his family did their own research, which produced a suspicion of the amoeba on their end. A few weeks later the official medical report came back, confirming Primary Amoebic Meningoencephalitis as the cause of death.

Officials closed the lake the summer of Waylon's death, reopening it the following year. Now, the family helps nonprofit organizations like Swim Above Water pass out flyers describing the amoeba in order to spread awareness, especially since Waylon's case is unique in that, unlike most victims described in the media, he was an adult.

Children Vs. Adults/Immune System

There are theories as to why children seem more susceptible to the amoeba, one being that kids are more active in the water than adults. Typically, adults swim at leisure or lounge on the shore while children are bundles of energy in the water, diving, cannonballing, tubing, dunking one another, and so on. Thus, the odds are higher of water, and the amoeba, being forced into the nose. Along with this, proper swimming is very much a learned behavior, and children have not yet had the time or experience to learn how to safeguard themselves. A child and adult may, side by side, engage in the same water activities, but an adult is experienced in holding his breath or otherwise not getting water up the nose, while a child is more likely to make a mistake and inhale water by accident.

Another theory is that, just like with viruses, the strand of amoeba contracted may have various strengths. One strand might be less virulent than another, which takes it longer to progress within the body, or makes it easier for the body to fight off.

And finally, a third viable theory is that adults have a better immune system built up, and so the body has a stronger chance of fighting the amoeba before it enters the brain. Immune system differences are certainly to be considered, Dr. Gompf agrees, along with genetic makeup. While there may be different strains of *Naegleria fowleri*, there is also likely a genetic factor that puts some people at risk and not others. After all, millions of people swim each year in warm water, and there are countless amoebas in that water, but only a select few are infected. There must be a reason why, and studies have shown that genetics may be a factor.

Research has been performed on mice in the past, where the mice are infected with *Naegleria fowleri*. But, only some of the mice showed signs of the infections, while others showed no negative effects. This is likely similar in humans. Many people may actually contract the amoeba, but for whatever genetic reason, the infection doesn't take hold.

One theory of genetic risk is simply how the amoeba travels through the nasal passages. During initial stages of a basic infection, the host (or, the person) secretes mucus that traps the foreign body. This process typically would stop the parasite from attaching to the tissue. *Naegierla fowleri*, however, "has a variety of virulence factors that may help the parasite to evade the host immune response and finally invade the tissue."[11] Simply put, the amoeba is able to evade the body's response to infection and push past the mucus barrier.

Once breaching the barrier, the amoeba must pass through cells in the nose, which it is able to do without destroying those cells — a fact somewhat specific to *Naegleria fowleri* alone. It is hypothesized that there may be bigger gaps in cells in the nose per person, which allow for the amoeba to pass through. Related to this, it has been shown that the cribriform plate in children is more porous than in adults, which would allow the amoeba to pass through more easily.

There may also be a difference in junctions between cells, or a personal genetic difference that makes one person more likely to contract the amoeba over another. Many micro-organisms, such as bacteria and viruses, "are able to disassemble TJ [tight junction] proteins and thus invade the host," while *Naegleria fowleri* is able to pass through without destroying cells, which may be another reason why it is so hard to diagnose and treat. There is evidence to suggest the amoeba may migrate via a parallel route, and that it degrades mucus by secreting a protease that allows the amoeba to evade the body's attempts to thwart infection.[12]

It likely takes a while for the amoeba to invade the epithelial layer (or, the cell layer that lines organs and glands, and that makes up the body's outer surface), but once it does breach that surface, the invasive population increases and continues to invade other tissue. In turn, this increases the likelihood that *Naegleria fowleri* will compromise tissues that are life-threatening for the host, such as the brain.

While it's unclear who is more at risk, or if genetics plays a role at all, one fact remains true: at this point in time, prevention is key for surviving *Naegleria fowleri*.

Chapter 8

Prevention

From the moment each victim became ill, and even more so upon diagnosis, the question arose — what could have been done to prevent this? With an infection that strikes so randomly, and so seemingly rarely by comparison to other illnesses, what steps could be taken to prevent contracting *Naegleria fowleri*?

Unfortunately, the only guaranteed way to prevent the infection is to avoid swimming or participating in any water-related activities, especially in warm freshwater. This seems an unlikely solution for many, however, and so for those who frequently (or even infrequently) go swimming, water skiing, tubing, etc., the best action to take is to limit the amount of water that goes directly up the nose.

For anyone who chooses to engage in water sports of any kind (and, let's face it, summertime in the south, especially, is spent in the water by most of us), some preventative measures can be taken. These include wearing nose plugs or holding your nose, keeping your head above the water, and avoiding jumping/diving/cannonballing into the water. It is also advised to avoid stirring up the sediment in shallow areas. The latter is advised because, although *Naegleria fowleri* thrives in warmer temperatures up to 115 degrees Fahrenheit, it can still grow and survive at colder temperatures for short periods of time.[7]

Once the amoeba is cultivated, it goes through three lifecycle stages: amoeboid trophozoites, flagellates, and cysts. Only in its first stage is the amoeba infective. Simply speaking, a trophozoite is the feeding stage of an organism, which makes it the most dangerous. This first stage enables the amoeba to not only reproduce, but also infect humans and animals via the nasal cavity, penetrating the tissue and migrating to the brain, where it causes PAM by literally feeding on brain tissue.

A flagellate is a non-amoeba-like dispersal form that allows an organism to "swim" and thereby spread until it encounters an appropriate resource, such as organic matter at the bottom of a pond. Once this resource is found, the flagellate transports into a trophic form, which is a larger and slower form that consumes matter using pseudopodia (finger-like projections that help move the amoeba toward its meal).

These trophozoites can be killed quickly by refrigeration, meaning they will die off once water temperatures drop if they are still in their first stage, but the amoeba in its third cyst stage can survive anywhere from weeks to months at cold temperatures above freezing where the environment is not conducive to feeding. The cyst is the dormant stage, protecting many types of organisms when food runs out, ponds dry up, or temperatures fall. Many small pond-dwelling organisms revert to this phase as a form of self-preservations until more favorable conditions return.[3]

This means that even in colder temperatures where *Naegleria fowleri* originally thrived during warmer months, the amoeba can still exist within the bottom sediments by converting itself into a cyst form that offers protection from the freezing water temperature.[3] A cyst form for aquatic organisms isn't uncommon, as it allows them to better handle temperature and water level stress.

In fact, these double-walled cysts can survive for upwards of eight months in waters as low as 39 degrees Fahrenheit (4 degrees Celsius). When water temperatures rise in warmer months, the amoeba is able to come out of its dormant cyst stage and revert back to the active trophozoites stage, and thereby reproduce. This means that potentially billions of people come in contact with and are exposed to *Naegleria fowleri* each year, but very few develop PAM. Those who do are typically from areas in the southern United States, though exposure has also been traced to central and northern US, south Australia, New Zealand, parts of Europe, Africa, Asia, and Latin America.

It is safe to say, then, that the amoeba is potentially a worldwide threat for men, women, and children across the globe who enjoy recreational freshwater activities. The median age for those who do contract the amoeba is approximately age twelve (78%), with 85% of these cases occurring in summer months in the southern US.[16].

With such a global environment and widespread death rate statistic, the need for better prevention methods is greater than ever. For the medical community, this means finding out more about the amoeba and discovering a cure. For the general public, this means being more aware of one's surroundings and taking proper precautions to ensure a safer swimming environment. Taking precautions can prevent individual deaths, and, perhaps more importantly, outbreaks of the amoeba.

In 2002, two Arizona children died within one day of another after coming in contact with the amoeba in untreated well water. PCR confirmation revealed *Naegleria fowleri* within the well water, which was most likely being used for bathing (in addition to drinking, but the amoeba is currently not shown to enter the body via drinking because stomach acid kills the amoeba, whereas these acids are not present in nasal passages). These deaths prompted more well water testing throughout Arizona, and, through various testing stages of water throughout drinking and plumbing systems, a startling 16% to 18% of the wells in the state tested positive for *Naegleria fowleri*.[16]

In 2004, an outbreak was reported in Czechoslovakia. There were sixteen cases of *Naegleria fowleri* within just three years, with the infection finally traced back to a poorly maintained swimming pool with a low chlorine content.

In 2012, more than twenty people in Pakistan died of the amoebic infection. The deaths are believed to be tied to ritualistic nasal cleansings, a fact that inspired the nation's government to advise on proper water treatments via chlorination.

2012 saw the deaths of two Louisiana adults due to the use of nasal rinses like neti pots. Typically used to help treat sinus infections, nasal rinses are designed to push water through the nasal passages in order to flush out excess mucus. However, in doing so, potential bacteria within the water is forced into the nasal cavity. In the case of the Louisiana victims, the amoeba was present within the water.

For this reason, the CDC recommends using only boiled, distilled, or filtered water whenever using a neti pot or related device. However, some studies have shown that even when using saline packets provided with the nasal devices, the amoeba isn't entirely killed off. During a study conducted by the CDC, saline solution packets were emptied into distilled water, with growth controls set up to measure results. Eighteen hours later, more than half of the amoeba had died, but that a significant population still thrived.[13] These results indicate that even when attempting to properly treat water for a nasal rinse, users may not be allowing the proper amount of time for all amoeba to die.

While the amoeba cannot live in high salinity in maritime environments, the smaller saline doses in nasal rinse treatments have essentially proven ineffective. Since these deaths, along with others, Louisiana has begun chlorinating all water purposed for residential and commercial uses to help prevent such infections. Though past tests have shown that adding various saline treatments were ineffective for immediate use, we can hope that these new treatment efforts will help save lives.

Make Ourselves Aware

The above are only examples of *Naegleria fowleri* infections that could have been prevented, either by ensuring water is properly treated or utilizing safer means of public health care. The question that remains, though, is whether anyone knew of this potential threat in the first place.

For so many of us, cannonballing into a pool, jumping in a puddle, using a nasal rinse device to clear out a sinus infection, all like millions have done before us, are actions so commonplace that the idea of prevention for a deadly brain-eating parasite never crosses our minds. And, for so many of us, the idea never crosses our minds because we were never even aware of such a threat, or that the danger of one tiny amoeba could be so close at hand, waiting to strike during an innocent outing in familiar waters.

Even Courtney had spent dozens of days past swimming in the river, hanging out with friends at local water parks, enjoying the summer in pools. Courtney's friends posted pictures to her Facebook wall just days after her death that commemorated her love for the water — as well as the day that would later become known as the one where she came in contact with the amoeba. In these pictures, Courtney is doing what all teens do: jumping off the dock into the water and showing off her back-flip skills. In two of the pictures, she is doing what specialists recommend – holding her nose before hitting the water. Even unknowingly, Courtney was protecting herself against the amoeba, but all it took was one time jumping in the water, one time being dunked by a friend, splashing around in the river, *without* holding her nose for one amoeba to enter her body.

So how do we prevent a silent killer, never knowing when it will strike? The first step is being aware of its presence, knowing where to potentially find traces of *Naegleria fowleri*. The second is taking proper measures to safeguard oneself from the amoeba, whether by boiling water for a nasal rinse device or wearing nose plugs when swimming. And the third step is doing our part to help others be aware.

We, as a whole community of concerned citizens, parents, brothers, sisters, spouses, must know what we are up against and demand answers. If we return to that daunting 86.2% of encephalitis deaths from 1989 to 1998 that are listed as unidentified, we can take it a step further. There were 11,200 deaths in that timeframe. If even one-fourth of the unidentified cases were due to *Naegleria fowleri*, that would be approximately 2,800 deaths — an average of three hundred and eleven per year. If one-sixteenth were due to the amoeba, then that would still average upwards of eighty deaths per year.

Of course, this is merely speculation on those unidentified deaths, but it does bring to light the possibility of a so-called "rare" disease not being so rare at all. The more awareness spread about this deadly infection, the greater chances we have of properly diagnosing *all* cases, and thereby fighting it. With awareness comes research, and with research comes more effective ways to test for and treat *Naegleria fowleri.*

Chapter 9

Testing Methods

With the rise of awareness organizations and the need to ensure proper prevention, so too do we, as the general public, need better ways to test for the amoeba — especially in early stages of diagnosis. The problem, however, is that there currently is no foolproof test method. While doctors do have ways to test for the amoeba's presence both when it is in the water and after it enters the body, these methods are not utilized until it is too late.

There are several ways currently to test for the amoeba. These ways include:

Direct Visualization

Naegleria fowleri is considered a motile amoeba, meaning that it is a single-cell organism capable of motion. Because of this, it can be spotted via a microscope in motion, often rapid motion, if it is a fresh sample. This is particularly helpful for both prevention and diagnosis, as it allows scientists to determine the presence of the amoeba in a body of water or doctors to make an early diagnosis and administer proper treatment.

Immunohistochemistry (IHC)

If a sample isn't fresh enough to allow for a motile amoeba, scientists can apply a stain directly to the amoeba within a tissue sample. To do so, an antibody that specifically treats *Naegleria fowleri* can be used alongside a secondary antibody designed to deposits a chemical that glows under IFA (Indirect Fluorescent Antibody).

Polymerase Chain Reaction (PCR)

PCR is a specialized biochemical technology that allows researchers to amplify one or several samples into millions. Often, this technology is used to produce copies of DNA or bacteria for molecular biology research. As an added benefit, copies can be made in a matter of hours for more timely results. PCR also allows scientists to determine if the amoeba is present in water or tissue samples, though because little is known about the amoeba's population within its natural environment, findings can be hard to interpret accurately.

Ameba Culture

A culture is used to study samples, whether to better understand the amoeba or detect its presence. For the purposes of testing, a sample is placed on a growth plate, then covered in a food source (in most cases, specific types of bacteria). *Naegleria fowleri* is then incubated at higher temperatures to encourage growth and allow researchers to study the amoeba as it travels across the plate and consumes the bacteria.

Environmental Detection

Perhaps one of the most important methods of testing is within the environment itself, as it allows for early detection and prevention. Within this method, water samples are collected and concentrated for testing via amoebic culture growth, then studied specifically for *Naegleria fowleri*.[8]

Together these forms of testing can save the lives of countless men, women, and children throughout the United States, as well as the rest of the world. First, though, they need to be made more readily available to ensure they are quickly employed.

Along with these methods, specific processes for testing should be followed. For example, amoebic samples (such as spinal tap fluid) should never be refrigerated, examined under high lighting, or placed in a high-speed centrifuge, as doing so may slow or damage the trophozoites. To best view a healthy motile amoeba, a CSF (Cerebral Spinal Fluid) gram stain using a simple slide and coverslip will help pathologists see activity. Misinterpreting this gram stain for an infection like meningitis means that any sort of antibacterial regime will be for naught in treating *Naegleria fowleri*.[17]

The above testing methods are only effective, after all, if they are utilized promptly. Otherwise, for the victims, it is too late.

Meeting Dalton

For young Dalton Counts, testing allowed for a faster diagnosis than Courtney, but treatment proved ineffective. Dalton and his family had gone swimming at Carpenter's Bluff in Texas on Sunday, July 22nd, 2012, a popular area where families get together to fish, swim, airboat, and enjoy various water sports. Summer weekends see crowds of thirty or more children with their families, as was the case the day Dalton enjoyed his time in the water.

By the following Wednesday afternoon Dalton began showing initial signs of illness, which at first mimicked signs of the flu. While staying home with his grandmother, he lost his appetite and developed a fever and vomiting within the next couple days. Thinking he was sick with a stomach bug, his family tried to get food in him by offering Sprite and crackers, but when his symptoms worsened, he was brought to the ER around eight PM.

Observation and testing began in the ER, where Dalton was initially treated for dehydration. Within three hours, he developed sudden blindness, which was realized immediately when he asked for the lights to be turned back on, when they had never been shut off. Doctors determined the cause of the blindness to be swelling on the brain, which later resulted in seizures. A spinal tap was administered at the Children's hospital in Dallas, and because Dalton was still having seizures, he was intubated so he wouldn't choke on fluids.

Saturday morning, around six AM, the official diagnosis of PAM was received. This diagnosis came a week after Dalton came in contact with the amoeba, as medical officials initially had trouble IDing *Naegleria fowleri*. But, with this diagnosis, doctors put Dalton on approximately fifteen different antibodies and inserted a catheter in his head to drain the fluids caused by swelling. The next Monday they began experimental treatments, administering the medications directly into the head catheter. This was a particularly risky treatment, as it could have made the swelling much worse, or maintained pressure while treating the infection.

At first, Dalton responded well to the treatments, but the next day, Tuesday morning, his pressure levels spiked. An MRI showed no brain activity, and doctors tried to bring Dalton out of the medically-induced coma. But, by 5:30, Dalton had passed away.

He was just nine years old.

Despite previous testing and treatments targeting *Naegleria fowleri*, Dalton's official death certificate lists swelling of the brain as the cause of death, with no mention of the amoeba or even meningitis. This 'official classification' mirrors so many others that list a similar disease, but not the actual cause.

Why is this?

In the wake of her son's death, Angel Laurentree began trying to answer that question. Like so many other parents who lost children to the brain-eating amoeba, Angel joined forces with local and distant awareness organizations. She has also participated in local poker rides (motorcycle runs) and rallies for awareness, utilizing her passion for motorcycles to help strengthen a great cause. Whether she's handing out flyers describing the warning signs of *Naegleria fowleri* or speaking to the public about her son, Angel is doing her part to educate everyone about the signs, symptoms, and treatments of the amoeba and PAM.

Children being Children

With the help of these organizations, she hopes to soon start up her own Texas foundation in Dalton's honor. Dalton loved to read, and even had a yearly subscription to *National Geographic* that, even at his young age, he looked forward to each month so he could read every new issue from cover to cover. Angel hopes to inspire that same love of reading, perhaps by offering books to underprivileged children. "Buying a book for a child who can't afford it is a gift from Dalton," she asserts.

Angel hopes that Dalton's story, his gifts, will help save the lives of children everywhere and prevent not only deaths, but also a mislabeled death certificate that hides the truth to a deadly illness. But with these gifts, she also hopes that kids get to stay just that — kids. "I worry about the fact that people are terrified to take their children swimming," Angel says, thinking over her plans for her foundation and what she hopes to teach people. While she wants to spread the message of awareness, she also believes that parents shouldn't let children stop being children.

"I wouldn't change a thing in the world," Angel continues. "[Dalton] was an awesome, smart, friendly, bright child and I have gone through the whole mourning phase of 'what if.' If I had done something else, he may not have had as much fun." Her message speaks of changing how her son was raised, letting him enjoy those moments of swimming and playing sports and reading beloved books, moments children know and love so well, rather than hiding him away out of fear of what might happen.

In fact, Angel carries this message with her at all times and applies it to her own life. Two weeks after Dalton's passing she returned to Carpenter's Bluff. Though she didn't get in the water, she did take the time to ponder such things, reflecting on her son, his illness, and his death. On the one-year anniversary of his death she again returned, and this time waded in. Some people may find it strange that she got in the same water that took her son's life, she acknowledges, but swimming in that very lake was something Dalton loved doing every week, and so she felt compelled to at least take one final day to swim in the same waters. "It was kind of weird, but relieving also," Angel recollects of that moment, remembering how she felt that day, a day when her beliefs of the importance of childhood fun perhaps meant the most.

"Don't take your child's life away because of what happened to someone else," she asks of parents, "but be cautious."

Dr. Rathore agrees with this, not wanting people to be afraid of swimming or to avoid freshwater locations, but rather to be aware and safeguard themselves against potential threats. Taking precautions saves lives, while still allowing children to be children. And, with each precaution taken, with each new person made aware of the dangers, we are one step closer to discovering more effective treatments.

Chapter 10

A Step toward Survival

With the rise in methods of testing, and better accuracy as a result, we can similarly hope to see a rise in survival statistics as well. To date, there are only three known survivors in 1978, 2003, and 2013, two from the United States (California and Arkansas) and one from Mexico. In all three cases, extensive medical treatment was required.

For the patient in Mexico, it took forty hours of hospitalization after admission to show any signs of improvement. However, it was twenty-two days after that a brain scan finally showed no signs of the amoeba or infection.

For the patient in California, it took a one-month hospitalization until no signs of PAM were detected. During this time, the patient was treated with a host of medications, including amphotericin B, miconazole/fluconazole/ketoconazole, and/or rifampin. These medications have been administered to other patients after contracting *Naegleria fowleri*, but with less fortunate results. Theories have been made that perhaps the strain of *Naegleria fowleri* within the California patient was less virulent than those of the fatal cases, thus aiding in recovery, but no concrete conclusion has been reached.[5]

Twelve-year-old Kali* of Arkansas is the most recent survivor in 2013. It is believed that Kali contracted the amoeba after a trip to a freshwater water park. After she began showing the traditional signs of the amoeba, she was admitted to the Arkansas Children's Hospital, where was placed on a breathing tube. Due to early detection, doctors were able to begin treating her quickly with more experimental types of testing, including a cocktail of antifungal drugs, antibiotics, and a new drug obtained directly from the CDC.

This drug, called miltefosine, was originally developed to help cure breast cancer. Together, these drugs helped treat PAM for the Arkansas child and, after nearly a month in the ICU, all traces of the amoeba were gone. Even after leaving the hospital, Kali still had to undergo physical and speech therapy to repair damaged caused by PAM. However, the fact that the treatment worked is promising, as the typical outlook for anyone who contracts the amoeba is poor, which is why early diagnosis and treatment are so critical.

Cases like Kali's are aiding scientists in breaking ground on treatment. If anything, doctors have learned that *Naegleria fowleri* needs to be treated like cancer, using three or four different drugs, all with different purposes, in order to kill the bacteria or infection. After all, cancer is treated using a myriad of medications, so why wouldn't we do the same for PAM?

Part of the problem is that the only drug proven to have a positive effect on the amoeba, miltefosine, isn't approved by the FDA for anything other than cancer. It also comes from Germany, which means it must be shipped to the United States whenever it is needed. In the past, there were no supplies kept in the US because the disease is so rare, so there was no need. By the time doctors determine a diagnosis of *Naegleria fowleri*, it is often too late to have the drug approved, ordered, and shipped overseas.

Just recently (as of 2013), the CDC received approval from the FDA to have its own stockpile of miltefosine, which means the drug can be released for medical treatments as soon as it's requested, thereby cutting time to ship from Germany and potentially saving the life of the patient in need.

In addition to being readily available, this stockpile of miltefosine allows for greater research to be done on *Naegleria fowleri* treatments via the Investigational New Drug Application through the FDA. The IND application is required for any new drug before it can be distributed across state (or country) lines, and allows for a sort of 'exemption' from the FDA so that official approval doesn't need to be made before the drug can be administered.

While going through the IND application process, the product at hand is tested for safe use in humans and to see how effective treatment is. The more data collected that identifies the drug as safe for use, with limited associated risks as shown via various clinical studies, the faster said drug will attain FDA approval.

In the case of miltefosine and the CDC obtaining its own stockpile, researchers can now track how the drug is used and the outcome. Data will show any and all medical situations where the drug is used, as well as whether or not it was effective. This information will be very helpful in the long run, and is vital in transforming the way *Naegleria fowleri* is treated.

It is also important to note that in addition to miltefosine, "the drug of choice for treatment of PAM has been Amphotericin B in combination with rifampin and other antifungal agents." Intravenous and oral administering of these medications have proven effective, but currently only miltefosine is in the research process for use in treating the amoeba.[10] The medication proved effective for Kali, but was also used to treat a young boy not long after her case, who did not survive. Additional testing will still be required in order to determine the best course of action.

Living on through Others

For the victims of *Naegleria fowleri*, their deaths have not been in vain. Every case has aided more information to the medical community, and some to the lives of others. For Courtney, being an organ donor was extremely important to her, and she had even updated her donor listing within days of contracting the amoeba.

Within twenty-four hours of her passing, an organ transplant team approached Courtney's mother, P.J., about utilizing her organs. Though she passed away from a severe brain infection, PAM had not shown to affect the other organs except the heart due to medications administered, and so many of her organs were able to be used for other patients.

Thus far, organs transplanted from donors infected by *Naegleria fowleri* have all tested negative for traces of the amoeba. However, the CDC has recently recorded some instances of the infection outside of the brain in two deceased victims of PAM. These infected sections included the lung, kidney, heart, spleen, and thyroid. This new data does suggest a risk of *Naegleria fowleri* transmission via organ donation, but additional study is needed to further explore the risks and benefits of organ or tissue donations.

For Courtney, this risk was one doctors were willing to take with her own organs. It was at this point that Amy*, an organ procurement coordinator at the hospital, began her journey with Courtney. Amy was at Courtney's side the morning doctors decided to pursue brain death testing, which showed that Courtney had lost all brain function due to the amoeba and no further treatment could save her. This official diagnosis was confirmed by three physicians assigned to her case over a certain length of time. Upon her death, the nursing staff notified a local organ procurement organization. Amy was one of the nurses on call that day, and it was then that she met the Nash and Ryder family.

During their first meeting, she explained her reasoning for being there, and that she was there to try to help them through this trying time. Having already looked through Courtney's file and being familiar with the teenager's organ donor status, Amy offered her family the opportunity to allow Courtney to potentially save several other lives through the gift of organ donation.

"I remember the hair standing up on my arms when I realized that this young woman had made these decisions for herself at such a young age and what a selfless act she was deciding for herself," Amy says, thinking back to the day she read through sixteen-year-old Courtney's case. "It brought tears to my eyes and thoughts to my head that maybe Courtney knew something none of us did. How could this young girl care so much about others to have registered online as an organ donor at the allowed age, then re-register when she received her driver's permit, only to update and reassure her wishes were again valid days prior to her death. All I could think was what an amazing girl this must be and I wished that I had known her."

This admiration and dedication to Courtney's case made the decision that much easier for P.J.

and her family. P.J. knew what her daughter would have wanted, and was happy to know that even after contracting such a deadly infection, that many of Courtney's organs could still be used. Research had been done prior to Courtney's case that proves the amoeba's effects are primarily contained within the central nervous system and do not cross over into the blood stream. Patients who receive organs are typically monitored closely over the course of several months, but thus far, no traces of the amoeba have ever been found in donated organs.

With this information, Courtney's family agreed to donate her organs. Doctors and nurses began the process of organ management throughout the evening and overnight. During organ maintenance and testing it was determined that Courtney's heart had sustained some damage, most likely from the stress of her illness, the medication, and what is called a "catecholamine storm" when the brain began to lose its functioning. She had also been on the ventilator for several days, so Amy spent some time optimizing her lung function. Lots of blood testing and tissue typing had to be done as well in order to save as many organs as possible for transplanting.

Ultimately, just one day after her death, Courtney was able to donate both lungs, her liver, both kidneys, and her pancreas. Through her gift of organ donation she saved the lives of four people (one patient received both lungs, another received both the pancreas and the left kidney, the third received the liver, and the fourth received the right kidney).

"Because of Courtney's generosity these people have a second chance at life," P.J. says of her daughter's organ donation. "It was amazing how people all over the world came together to pray, give their love and support for her and her families. They held vigils praying for a miracle. We did not get our miracle, but Courtney's persistence to be an organ donor gave miracles to other families."

After working alongside the organ donation team, Amy joined Courtney's family in their efforts to raise awareness. In her own words, Amy says, "This family has included me in every event honoring Courtney and bringing awareness to the amoeba and to organ donation. I continue to be amazed by the strength that Courtney has brought to this family and their courage to continue to tell her story and to include myself as part of it.

"Over the two years I feel like I know this family as if they've been in my life as long as I've been here. As for Courtney, I never had the opportunity to know her here, but I have been extremely blessed to have been a part of her story. It has been an honor and amazing part of my life to be able to honor her and tell her story from my point of view.

"I believe Courtney was out in this earth for a very specific reason, to bring joy to everyone she met, and it is apparent that even in her death she has done that. I count it an honor and a pleasure to tell anyone who asks or doesn't about this very special angel, Courtney Nash."

Organ Donation Risks

While some organs have been successfully transplanted with no traces of the amoeba, there is still a risk associated with doing so that prevents many organ donors from being able to fulfill that wish.

Nine-year-old Christian Strickland of Virginia was at a fishing camp for the second year in a row when he came in contact with *Naegleria fowleri*.

It is believed that he contracted the amoeba after roughhousing with other boys and dunking one another in the water, as two weeks later he began showing signs of illness. Christian passed away on August 5th, 2011, to what doctors initially suspected as meningitis and encephalitis — though this could not be officially confirmed.

At the time, Christian's parents decided to donate his organs, and were disappointed to learn that doctors would not accept the transplant because his cause of death was unknown. It wasn't until the following weekend that the health department confirmed his death due to *Naegleria fowleri*.

The concern of using organs from patients who died from meningitis-type diseases is growing, particularly in cases with the amoeba. Recent autopsy reviews in the *World Journal of Transplantation* have shown cases of *Naegleria fowleri* in organs other than brain. These reviews focus on current and previous cases, finding evidence of the amoeba in the heart, lungs, spleen, and other organs that previously didn't show traces. Determining new traces of the amoeba suggests that it is spreading from the brain to the rest of the body — or that, perhaps, medical specialists simply weren't looking anywhere other than the brain before. Autopsies typically focused on the tissues and microscopic evaluations of brain, not other organs.

"You don't find what you don't look for," Dr. Gompf states matter-of-factly, referring to autopsies and reiterating the fact that testing may not have been done in the past to look for the amoeba in other organs. Once pathologists or doctors find what they are looking for in a test or autopsy, they can usually stop looking for anything else.

Today, autopsies aren't as common as they used to be. Whereas in the past one was performed for nearly any death, nowadays they are primarily performed in criminal cases, as this is the most standard realm of interest. Often, families have to hire a private pathologist or hospital volunteer in order to have an autopsy performed; otherwise, doctors may have to make a best guess based on what is known from symptoms or tests.

In fact, death related to an amoeba is not often listed on any death certificate, even in cases *proven* to be related to *Naegleria fowleri*. Many list meningitis or encephalitis, or some by-factor of PAM, such as heart failure or swelling of the brain. Because of this, *Naegleria fowleri* may appear rarer than it actually is, which leads to the CDC having a harder time acquiring life-saving drugs needed to treat patients as soon as possible. This is why recent autopsy reviews have proven so valuable both to the medical community and public as a whole, as they prove the need for more in-depth testing while also shedding light on the amoeba.

For Philip Thomas Gompf of Florida, these autopsy reviews gave his family some amount of closure needed regarding their little boy's death.

Philip was a passionate little boy with a big sense of humor and even bigger desire to change the world and make things right for people. An avid reader, Beatles fan, and aspiring writer, he dreamed of becoming a scientist or engineer. He loved taking things apart and rebuilding them, especially his father's old circuit boards.

Just days after his ninth birthday, on a Saturday, Philip went wakeboarding with his cousins on Lake Arietta in Polk County, Florida. This fun-filled weekend was followed soon by symptoms of *Naegleria fowleri*, which appeared on the Thursday after the previous weekend. The symptoms began with headaches, a fever, and a stiff neck. His father, a pediatrician, knew these symptoms to be serious and rushed the boy to the hospital. A spinal tap was performed and came back showing signs of a severe infection. It was at that point that Philip's father and mother, Sandra Gompf — both doctors educated in the dangers of meningitis-type infections — knew he wouldn't make it.

As expected, Philip's condition worsened at shocking speed, transforming to hallucinations and seizures within hours of the spinal tap. He was helicoptered to Tampa General, and once there went into heart failure. A neurosurgeon was called in to check for and potentially treat swelling on the brain, but by then it was too late. Within just five days of showing initial signs of the infection, Philip passed away on August 17th, 2009.

His official cause of death was listed as heart failure caused by meningitis.

As has become more common, the medical examiner declined to perform an autopsy, so the Gompfs requested one and hired a private pathologist. The diagnosis of *Naegleria fowleri*, not heart failure caused by meningitis — arrived a month later. Despite the diagnosis, the Gompfs were able to donate Philip's kidneys, saving the lives of two recipients.

However, they were not able to donate his heart. Initially, not having known of his diagnosis, this was a difficult fact to accept. But, the recent studies in transplant donors in the journal were enlightening for Philip's family, as they showed traces of the amoeba in Philip's heart. Prior to the reviews, doctors thought his heart failure, which appeared during his last few days in the hospital, was caused by the stress of the situation. His parents, though, disagreed, believing something else to be the cause, as heart failure wasn't typically a symptom of the amoeba. The new autopsy reviews answered that question for the Gompfs, while also reaffirming the need for more thorough testing in order to ensure more organs aren't being affected, and that those organs aren't being transplanted.

Chapter 11

Make a Change

To this end, *Naegleria fowleri* must be tested for immediately after patients are admitted to the hospital as well as in autopsies. Meningitis and encephalitis are so broad in symptoms that they are easy to be listed as the "go to" for diagnosis. PAM is so rare, and so rarely talked about, that it's not at the top of the list in terms of testing and diagnoses. 99% of time, doctors will look to the former based on patients' symptoms. "Most physicians will go their entire career without ever seeing [*Naegleria fowleri*]," Dr. Rathore states. Many don't have experience with the condition, and so both doctors and family members alike simply aren't aware that they should be looking for the amoeba, or requesting testing specifically for it. "One day [patients] are fine and the next thing you know they are sick, and parents aren't sure what to do," Dr. Rathore explains. Parents or spouses don't think of an amoeba first, they think of the flu or a stomach bug. And in thinking this, they don't even consider an alternative infection. "We have to suspect it, first of all. What the brain does not know, the eye will not see."

"People need to ask about it," Dr. Gompf echoes Dr. Rathore's statement, speaking from experience regarding patients and their medical care. "This means more than just reading something from the CDC. The more parents actually ask, the more likely doctors are to consider it. This means a lot to doctors. They want to please, find the cause just as much as you do."

In a sense, diagnosing *Naegleria fowleri* is very much a matter of taking treatment into your own hands. As with any infection, the sooner you treat it, the better the outcome. But one of the biggest issues with the amoeba is that it must be diagnosed within a matter of hours, and treatment needs to begin within twelve hours of the first day of exposure. For most cases, symptoms don't even show until at least a day after contracting the amoeba, which automatically lowers the chances of treatments working.

So, it is up to everyone individually to know and accept that coming in contact with *Naegleria fowleri* is possible. As soon as an illness worse than the flu is suspected, it needs to be brought to a physician's attention that the patient was swimming and the amoeba may be to blame for the sudden onslaught of symptoms. All infections are selective and can take a long time to diagnose, and we (the collective "we") still don't know exactly how effective current treatments are. All of this together makes treatment even more difficult, because only by diagnosing and treating the infection can more research be done. In order to find a good treatment that actually works, changes need to be made on many levels, and PAM is so rarely diagnosed that any change is not easy to do. Asking, or even demanding, that a physician check for PAM is a step in the right direction.

But it is more than that. Having the awareness of PAM means speaking up, telling doctors to look for the amoeba if you suspect it. Don't be afraid to speak up, even if it's as simple as suggesting to a friend that they use distilled water rather than tap water for a nasal rinse, or as complicated as ordering a doctor to check specifically for an amoebic infection.

Together, everyone can bring about the awareness needed to change — and save — lives.

Chapter 12

Families Left Behind

In the aftermath of each person's passing after contracting *Naegleria fowleri* comes the question of survival for those left behind. How do families, friends, move on from such a sudden and fatal attack? Just as Jim Ariola expressed, we become different people in our grief, and moving on, returning to some semblance of a normal life, is no simple matter. For many, moving on meant joining a cause, and ensuring their loved one is never forgotten.

Even two years later, Courtney's friends and family members continue to post on her Facebook page, sending thoughts of goodwill to their late friend. A simple 'missing you today,' an anecdote of a particularly humorous event, a happy birthday or Merry Christmas. To honor Courtney's love of the Florida Gators, a brick with her name was placed at the University of Florida. At Christmas, her family puts her initials in lights, and throughout the year, the community comes together to remember the departed. Courtney's school likewise gathered funds, with generous grant money from Lowe's as well, to create the Courtney Jean Nash Outdoor Learning Pavilion on her school campus. Students and faculty alike came together to lay bricks and plant gardens that would yield assorted fruits and vegetables, which will be used to help others in need.

More than messages and lights, though, comes

the drive to raise awareness. In the wake of a loved one's death as a result of PAM, families are left with the desire, the need, to figure out the cause and treatment, to raise awareness about the amoeba and resulting infection that everyone should know about. For Courtney's family, this need has taken on several forms.

Together, P.J. and her family started the Courtney Nash Amoeba Awareness Foundation, Inc., an organization dedicated to educating the public while telling Courtney's story. Through the Foundation, numerous events have been held in Florida to help save the lives of countless individuals.

Just one example of these events is the Memorial Blood Drive held in Titusville on July 13th, 2013. This was one of two blood drives held to date, with more planned throughout the coming years in partnership with local organizations that provided donations. These blood drives are a way to give back to the community, educate as many as possible, and raise awareness on this deadly amoeba. Free nose plugs are given out along with bumper stickers and license plates that warn of the deadly killer. The Foundation has also started up a scholarship program, titled The Courtney Jean Nash Memorial Scholarship, which provides financial assistance to high school students heading off to college.

In addition to these events, which are highlighted by the local media to reach a wider audience, the Foundation has created banners and warning signs to complement the license plates and bumper stickers, warning people of the dangers of swimming in warm waters. Because Florida is a tourist state filled with rivers and lakes, the need for these signs is greater than ever, particularly in spring and summer months, which is why banners are placed along I-95 and at local schools and stores. Together, these awareness efforts reach more people and garner the attention needed to engage the public and educate everyone who sees them.

Above all, the Courtney Nash Amoeba Awareness Foundation operates with the belief of being proactive to find a cure and put an end to deaths by *Naegleria fowleri*. The Foundation holds the goals of:

> **1)** Helping people realize that any body of water, fresh or brackish, that reaches a temperature of 80 degrees Fahrenheit or more is a potential hazard.
> **2)** Gaining funds that are needed for additional research to try to find a cure and create awareness to prevent this from happening to anyone else.
> **3)** Ensuring physicians are more aware in order to accurately and rapidly diagnose this killing parasite.
> **4)** Teaching others about organ donation so that other lives may be saved.

P.J. and her family always seek to extend a thank you to the doctors, nurses, and staff at Arnold Palmer Hospital for their exhaustive efforts taken to try and save Courtney. They especially want to give credit and praise to the infectious disease specialist who volunteered his time to take on a case not originally his own in order to help save the girl he never had the chance to know.

Arnold Palmer is a teaching hospital, and one that goes beyond the standard doctor-patient relationship in order to work together and combat life-threatening diseases. Currently, those same doctors and staff are putting together a task force in order to better diagnose, treat, and destroy this horrible disease caused by an innocent day of swimming. Their hard work and dedication will never be forgotten by the Nash or Ryder families, and P.J. is always the first to praise the medical team that never left her daughter's side.

Swim Above Water

The Courtney Nash Amoeba Awareness Foundation isn't the only organization dedicated to such causes. Swim Above Water, headquartered in Minnesota, is likewise devoted to the mission of creating "awareness through education, blogs, and specialty events throughout Minnesota and the Northern Regions." This organization seeks to teach about the dangers of swimming in freshwater systems and how to prevent infection of the amoeba *Naegleria fowleri*.

Swim Above Water was founded in 2012 by friends of Jim Ariola, after the death of his son, Jack Ariola Erenberg. About six months later, Chad and Heidi LaMeyer joined. Now, Jim and Heidi make up the board, and Chad serves as treasurer. Together, the board members and those who contribute to Swim Above Water work with the ultimate goal of saving lives.

"That is a difficult thing to measure," Heidi comments on her foundation's mission, "because you just don't know if your efforts can be documented in that nature. What we can do that is tangible is raise dollars to donate to research, provide education and information on prevention and also treatment concerns. We honor the victims of this disease as a platform for our other activities."

Since its inception, the organization has told their own story and reached out to others who lost loved ones, as well as been the host of awareness benefits and events, including their largest event *Awareness Rocks*, which brought in upwards of two hundred people. This event included a silent auction, drawings, and live music, with sponsorship opportunities for those interested in being a part of the organization. This annual event is already gaining popularity, and a Minnesota favorite musical entertainer, GB Leighton, has already signed on to be part of raising awareness.

Because the amoeba season (or, the season with warmer waters) is shorter in the north, Swim Above Water holds two to three events a year. Previous events have included talks at local schools, booths at community fairs, and awareness setups at art shows, with future events including a golf tournament. A related blog is also updated with stories of *Naegleria fowleri* victims, with the goal of building relationships with families and offering more large-scale awareness efforts. Swim Above Water welcomes partnerships and collaborations with other organizations, networking with foundations across the US whenever possible. With each event, dollars are raised to fund research for the CDC and to support awareness efforts now and in the future.

Swim Above Water works closely with the CDC as part of their amoeba division to help with awareness efforts. They are also part of the National Advocates for Amoeba Awareness, an advocacy group that is recognized by the CDC.

Above all, the common root of all events and benefits is education — education of the amoeba to both the public and medical community, of the need for more research and medical treatments, and of the knowledge that PAM can happen anywhere.

Swim Above Water primarily focuses on northern areas for this very reason. "People just don't believe it can happen," Jack's father, Jim, says. "They think it's a fluke and it can't happen to their kids. I know firsthand that it can."

Other families hold true to this statement as well, no longer letting themselves believe that it won't happen to them. For the Abel family, swimming in freshwater is no longer an option. Since Waylon's death, neither his children nor his immediate family have been in freshwater, as they are not willing to take the chance. Only his father returns to the lake that took his life, and that is to check water temperatures.

Acceptance of this message is vital, which is why so many families and organizations work as valiantly as they do. For, in order to ensure more survivors, the message must be clear that even in the coldest or warmest of climates, *Naegleria fowleri* can still thrive.

Chapter 13

Changing Lives

Courtney's story, as with the many others, has already changed lives and made a difference. One of these differences was inspired by someone people may least expect — an eleven-year-old girl named Jayde* of Titusville, Florida.

After completing an assignment at school, Jayde was permitted to go online and play math games on the computer. After signing on, she found what was meant to be a fun and lighthearted game called "Amoeba's Life," which centered on an amoeba that traveled around eating cells within the body. Having heard about Courtney Nash's death only a couple years before, and knowing about the devastating effects of this brain-eating parasite, Jayde was disappointed to see a game making light of such a heartbreaking disease. And so, the child decided to do something about it.

The game upset Jayde to the point of tears, tears for the pain brought to the Nash family by the amoeba itself, and tears over a game that she felt she was powerless to change. After all, what could an eleven-year-old do? Her first idea was to write a letter to whomever she could to petition a name change for the game. She expressed her thoughts to her father, who immediately went to the game's website and found contact information.

With his direction, Jayde wrote an email to the

game company, informing them about the amoeba and about Courtney, even sending them the URL for Courtney's foundation. With her father's help, she sent the email directly to the company's CEO, not knowing if anyone would read her email, let alone do anything about it, but happy that she was, at the very least, able to share her opinion on the matter. Her email was a start, and she hoped it would make a difference.

Within twenty-four hours, the CEO himself responded to Jayde's email, much to her and her father's surprise and delight.

In the email, the CEO apologized for the name of the game, stating he was unaware of the real amoeba and of the many lives it had claimed. His goal in creating the game was to help children learn while having fun, not to remind anyone of the heartaches families went through. And so, he agreed to change the name of the game to "Cootie's Life," taking away the relation to the amoeba while still having a game that allowed children to learn. The CEO also commended Jayde on bringing awareness to the fact that amoebas are very real, and not something to be made into fun games.

Despite her young age, Jayde recognized the problem with the game and sought to do something about it. At eleven years old, she educated an adult on the dangers that are alive and living in warm waters, and her story was shared with countless others as a result, thereby spreading awareness from one corner of the nation to the other.

Broadening Our Knowledge

Just as Jayde spread awareness to a computer

game creator, so, too, do those involved with various organizations. With each blood drive, rally, interview, school talk, education is given to fresh minds that are receptive to learning about this deadly disease. But their efforts aren't restricted to events around town or at local schools.

For some, awareness came through the writing of a book, a book that began by telling one girl's story and ended up sharing the lives of others affected by this deadly disease. Awareness, for me, meant forcing myself to become a journalist, a scientist, and a friend, a listener, adapting my way of writing and learning in order to better understand a world I knew nothing about. And this journey has opened my eyes to a disease that terrifies and fascinates me, and one that I now spread the word about in order to ensure those I love – and the ones they love – know what precautions to take.

The more people who know about *Naegleria fowleri*, whether they read a book or donate blood or see an episode of *Monsters Inside Me* (a television show that has featured a segment on the amoeba), and the greater chance we have of fighting this deadly parasite. And the more chances we have of fighting it, the more lives doctors will be able to save. Today is the day of technology, where social media reigns supreme when it comes to getting a message out, which gives everyone an enormous platform on which to tell their story. Sharing an article about the amoeba, discussing a book, writing a blog post that reaches thousands — the opportunities are endless to spread the word.

It is important to note, though, that with the rise in social media too comes the wealth of information about any given topic. As Dr. Rathore points out, "Information is not necessarily knowledge." As a nonprofit organization, journalist, author, or simply a concerned citizen, everyone must do their part to spread the word, and spread the *right* word. We do not want to incite fear or create sensationalist media, nor do we want to give false information to prove a point. Social media must be used responsibly, by everyone involved.

For Swim Above Water, social media has played a big part in their awareness efforts. The organization has made ground through their in-person events, which has been a bit of an uphill battle from the beginning, considering their location and the many Minnesota lakes are what draw and keep people there. Swimming is the state's livelihood due to tourism. But, while they are continuously growing in their event support, their social media presence is playing a big part in the organization.

To date the Swim Above Water blog receives thousands upon thousands of views, both by repeat visitors and new reaches. Heidi, the blog's host and author, has even been approached by people when out in public who have read the blog and want to offer their own support. What may seem like such a small thing, to offer words of praise or even donations to the cause, proves the larger-scale effects that any type of awareness has on the general public.

I have seen entire communities come together in the wake of a child's death, friends of victims gathering for celebrations of remembrance, families working together to tell their children's stories, strangers reaching out to help those in need. Each act of public education is a step closer to a cure, a step closer to changing the lives of those who may contract the deadly amoeba, and of those who are working so hard to make sure no one else loses their life to *Naegleria fowleri.*

I believe I speak for everyone involved in the

writing of this book when I say that our goal is to educate and prevent, and hope that readers take away the message that we all should be aware of what lurks in our warm freshwater systems, but not to let that fear hold us back from living our lives. And yet, so too should we join in on efforts to change lives.

Donate blood. Pass out flyers. Buy nose plugs before that family trip to the lake. Take a break from sitcoms to watch a scientific program on single-cell organisms that invade bodies.

And, most of all, take a moment to celebrate the lives of those lost to *Naegleria fowleri*, to send the message that the world is fighting back for them, so that no longer will anyone fear the single-cell danger that may threaten us during the most innocent of days, and a single swim.

In Memory Of:

Courtney Jean Nash
July 16, 1995 - August 13, 2011

Hailee LaMeyer
September 23, 1996 - August 23, 2008

Jack Ariola
January 19, 2003 - August 6, 2012

Will Mathews
May 27, 1996 - Aug. 15, 2010

Philip Gompf
Aug. 5, 1999 - Aug. 17, 2009

Christian Strickland
May 3, 2002 - August 5, 2011

Waylon Abel
August 12, 1981 - August 7, 2012

Mark Kinkade
September 19, 1963 - July 31, 1991

A Message for Courtney

My memories of my sister Courtney Nash will forever be in my heart. We were best friends. I could tell her everything. She loved everyone and enjoyed her life to the fullest. I hope I can be half the person she was. I will continue to share Courtney's story to warn others to be safe and protect themselves. I love you Courtney and miss you every day.

Love Ethan (Bubba)

A Special Thanks

Words will never be able to let everyone know how proud I am to have been a part of raising Courtney for thirteen years. Courtney was a perfect reflection of P.J. They are the most selfless people I know.

Being just sixteen, Courtney had a vision of what she wanted in her life, and was willing to do whatever it took to achieve the things she wanted.

My name is Brett Ryder. I am the president of the Courtney Nash Amoeba Awareness Foundation. Along with P.J., vice president, and Thomas Uzel and Christine Dugan (directors), we have dedicated and committed our lives and time to educating every person we can about the dangers of this deadly amoeba.

On behalf of the foundation, I would like to thank everyone from our hometown of Titusville, Florida, and everyone from around the world who have helped support our mission and honor Courtney.

Brett Ryder

Sources:

(1) Centers for Disease Control and Prevention. "Naegleria fowleri - Primary Amebic Meningoencephalitis (PAM)." <http://www.cdc.gov/parasites/naegleria/index.html> November 14, 2013.

(2) Manual of Environmental Microbiology. 3rd Edition. Edited by Hurst, Crawford, Garland, Lipson, Mills, Stetzenbach. American Society for Microbiology Press, 2007.

(3) Centers for Disease Control and Prevention. "Pathogen & Environment." <http://www.cdc.gov/parasites/naegleria/pathogen.html> November 14, 2013.

(4) Centers for Disease Control and Prevention. "Sources of Infection & Risk Factors." <http://www.cdc.gov/parasites/naegleria/infection-sources.html> November 14, 2013.

(5) Centers for Disease Control and Prevention. "Illness & Symptoms." <http://www.cdc.gov/parasites/naegleria/illness.html> March 5, 2014.

(6) Centers for Disease Control and Prevention. "Policy & Recommendations." <http://www.cdc.gov/parasites/naegleria/policy.html> December 12, 2013.

(7) Centers for Disease Control and Prevention. "Swimming." <http://www.cdc.gov/parasites/naegleria/swimming.html> November 6, 2013.

(8) Centers for Disease Control and Prevention. "Diagnosis & Detection." <http://www.cdc.gov/parasites/naegleria/diagnosis.html> February 12, 2014.

(9) Amoeba Response Protocol. Government of Western Australia Department of Health. <http://www.public.health.wa.gov.au/cproot/3649/2/Amoeba%20Response%20Protocol%20v3.pdf> 2013.

(10) Cabral, Guy A. & Marciano-Cabral, Francine. The immune response to Naegleria fowleri amebae and pathogenesis of infection. FEMS Immunoi Med Micribiol. (2007)

(11) Shibayama, Mineko. "Mucins in the host defence against Naegleria fowleri and micunolytic activity as a possible means of evasion." Microbiology, 2008.

(12) Serrano-Luna, Jesus. "Disruption of MDCK cell tight junctions by the free-living amoeba Naegleria fowleri." Microbiology, 2013.

(13) Yoder, Jonathan S. "Primary Amebic Meningoencephalitis Deaths Associated With Sinus Ir-

rigation Using Contaminated Tap Water." Deaths Caused by Naegleria fowleri Infections, 2012.

(14) Gompf, Sandra G., MD. "'Rare' Is Only Half the Amoeba Story." Personal Communication. 2014

(15) US National Library of Medicine. "Encephalitis-associated deaths by category and study period, United States, 1979–1998" <http://www.ncbi.nlm.nih.gov/pmc/articles/PMC2870605/table/tab001/> 2006.

(16) Mandell, Gerald L. Principles and Practice of Infectious Diseases. 7th Edition. Elsevier, Inc., Philadelphia, PA. 2010.

(17) Garcia, Lynne. Practical Guide to Diagnostic Parasitology, 2nd Edition. ASM Press, Washington, DC. 2009.

(Other) Heggie, Travis W. Swimming with Death: Naegleria fowleri infections in recreational waters. Travel Medicine and Infectious Disease (2010).

About the Author

Night owl, Dorito lover, and quiet eccentric - Kristina Circelli is the author of several fiction novels, including The Helping Hands series, The Whisper Legacy, The Never, and Fragile Creatures. A Single Swim is her first non-fiction book.

A descendant of the Cherokee nation and niece of a Cherokee elder, Circelli holds both a Bachelor of Arts and Master of Arts in English from the University of North Florida, where she teaches creative writing. She also heads Red Road Editing, a full-service editing company for independent authors and commercial clients.

She currently resides in Jacksonville, Florida with her husband, Seth, and cats, Lord Finnegin the Fierce and Mr. Malachi the Mighty.

Connect with Kristina

Website: www.circelli.info

Facebook: https://www.facebook.com/AuthorKristinaCircelli

Twitter: @KCircelli

Email: Kristina@circelli.info

Connect with Awareness Organizations

Courtney Nash Amoeba Awareness Foundation:
http://courtneynash.com/Home_Page.php
http://www.courtneynashamoebaawarenessfoundation.com/

Swim Above Water:
http://swimabovewater.org/

Swim Above Water blog:
http://swimabovewaterusa.blogspot.com/2014/04/2014-awareness-season-begins.html

Nationally Notifiable Diseases Info:
http://www.gompfsidpearls.net/Amoeba-Season/put-pam-on-the-national-notifiable-diseases-surveillance-list/

Printed in Dunstable, United Kingdom

64753596R00068